Explanation of Frontispiece
Photograph of a portion of the Great Wall of China where it crosses the Niuxinshan gold mining district (see text Fig. 3.1). The view is taken from atop the Niuxinshan granite looking west. The prominent cliff in the middle ground is formed by Late Proterozoic quartzites of the Changcheng (meaning Great Wall) System which overlie Archean metabasites along a thrust fault at the base of the cliff (fault F3 in text Fig. 3.2). Gold is being mined from quartz veins in the Archean rocks on the northeastern side of the Niuxinshan granite, about 1000 m from the site of this photograph.

R.B. Trumbull G. Morteani
Z.L. Li H.S. Bai

Gold Metallogeny
in the Sino-Korean Platform

Examples from Hebei Province, NE China

With 49 Figures and 21 Tables

Springer-Verlag
Berlin Heidelberg New York
London Paris Tokyo
Hong Kong Barcelona
Budapest

Dr. Robert B. Trumbull
Prof. Dr. Giulio Morteani
Lehrstuhl für Angewandte Mineralogie und Geochemie
Technische Universität München
Lichtenbergstraße 4
W-8046 Garching, FRG

Li Zhiliang
Bai Hongsheng
The First Geological Exploration Bureau
Ministry of Metallurgical Industry
Yanjiao, Eastern Beijing
People's Republic of China

ISBN 3-540-55231-6 Springer-Verlag Berlin Heidelberg New York
ISBN 0-387-55231-6 Springer-Verlag New York Berlin Heidelberg

Library of Congress Cataloging-in-Publication Data
Gold metallogeny in the Sino-Korean platform: examples from Hebei
Province, NE China/R.B. Trumbull... [et al.]. p. cm. Includes bibliographical references and index.
ISBN 0-387-55231-6 (U.S.)
1. Gold ores-China-Hebei Province. 2. Metallogeny-China-Hebei Province. I. Trumbull, R.B. (Robert B.)
QE390.2.G65G655 1992 553.4'1'0951152-dc20 92-17971

This work is subject to copyright. All rights are reserved, whether the whole or part of the material is concerned, specifically the rights of translation, reprinting, reuse of illustrations, recitation, broadcasting, reproduction on microfilm or in any other way, and storage in data banks. Duplication of this publication or parts thereof is permitted only under the provisions of the German Copyright Law of September 9, 1965, in its current version, and permission for use must always be obtained from Springer-Verlag. Violations are liable for prosecution under the German Copyright Law.

© Springer-Verlag Berlin Heidelberg 1992
Printed in Germany

The use of general descriptive names, registered names, trademarks, etc. in this publication does not imply, even in the absence of a specific statement, that such names are exempt from the relevant protective laws and regulations and therefore free for general use.

Typesetting: Best-set, Hong Kong
32/3130-5 4 3 2 1 0 – Printed on acid-free paper

Preface

This monograph reports the results of a cooperative research project conducted from 1987 to 1989 on selected gold deposits in eastern Hebei province, People's Republic of China. The European partners were the Technical University of Munich, Federal Republic of Germany, and the University of Trento, Italy. The Chinese partner was the Ministry of Metallurgical Industry (MMI) in Beijing, which is responsible for the development and operation of all major gold mines in China. In addition to data and observations collected during the project investigations, this monograph includes a great deal of information taken from published and unpublished geologic reports and maps from sources in China.

This research has been realized with financial assistance from the Commission of the European Communities under the EC International Scientific Cooperation Programme with the People's Republic of China. The views herein expressed are those of the participating scientists and do not represent any official view of the EC or of the government of the People's Republic of China.

Questions of Style

We have tried to develop a consistent style in this text while combining information and ideas from both western and Chinese sources. For the writing of Chinese names and localities we use the Pinyin system of transliteration and the following conventions:

a) All Chinese place names are written as one word; thus we write "Niuxinshan" and not "Niu Xin Shan". In the case of personal names, the family name is written first followed by the given name, and the latter is written as one word (e.g., Sun Dazhong).
b) Citations of both Chinese and Western authors are made in the same conventional style. For the Chinese authors,

only the family names are written in full, and given names are represented by one or two initials depending on the number of syllables in the name. For example, Sun Dazhong is cited as Sun D Z, and Li Ren is cited as Li R.

A troublesome problem of style concerns the nomenclature of the Early Precambrian rock units and the terms used to describe the regional tectonics. The Chinese geologic terminology does not conform with modern international practice in naming high-grade metamorphic terranes and it does not reflect modern concepts of plate tectonics. The authors of this book acknowledge this problem but they have neither the inclination nor the competence to update the nomenclature of Chinese regional geology. We therefore follow the nomenclature given in the recent comprehensive references to the geology of China (Yang et al. 1986; Ren et al. 1987). This may irritate some readers, as it has irritated one reviewer of the text, but it has the important advantage that one can easily look up details about specific lithologic or tectonic units cited here in the Chinese literature without having to guess at the original form of their names.

Acknowledgments

The work reported in this text would have been impossible without the financial support of both the European Community and the Chinese Ministry of Metallurgical Industry. The authors gratefully acknowledge this support. The European contribution to this project owes much to the work of Dr. Gerhard Lehrberger, Technical University of Munich, who helped with all field work in China and with a large part of the petrographic and geochemical studies in Munich. Prof. Muharrem Satir, now head of the Institute for Geochemistry, University of Tübingen, directed the isotopic studies of the European partners, and aided in field sampling for age determinations. His successor at the Technical University of Munich, Dr. Dominique Blamart, performed many of the stable isotopic analyses reported here, and made useful comments on this manuscript. We thank Prof. Andreas Fuganti and Dr. Guido Ceri of Geoexpert International, Trento, Italy for their work on LANDSAT satellite interpretations. Dr. Dietrich Ackermand, University of Kiel, gave access to his microprobe laboratory and helped with the microprobe analyses. Finally, the European partners wish to personally thank the staff members of the Ministry of Metallurgical Industry (MMI) for the warm hospitality and efficient organization which made our stays in China a pleasure. In particular we wish to thank Mr. Yao Peihui, Mrs. Hu Pinmei, and Mr. Lu Jin of the Geological Bureau of MMI in Beijing; and Mr. Li Maocai, director of the First Geological Prospecting Bureau of MMI in Yanjiao, for their untiring support of our joint project. A final word of thanks is due Mr. Liu Hua of the First Geological Prospecting Bureau of MMI for his excellent service as translator during our stays in China.

The authors from the Ministry of Metallurgical Industry wish to thank the Analysis Center, Chinese Academy of Geosciences of the Ministry of Geology and Mineral Resources, and the Analysis Center of the Geoscience Institute of Academica Sinica for support in providing chemical analyses and interpretations. For their help in field investigations at the various deposits we thank Mr. Ma Wenrong, engineer of the No. 515 Geological Prospecting Team, First Geological Prospecting Bureau of MMI, for support of investigations in the Niuxinshan district; and Mr. Yuan Yingliang, engineer of the No. 522 Geological Prospecting Team, First Geological Prospecting Bureau of MMI, for guidance in the Sanjia district. For his excellent petrographic work on rock and ore sections we thank Mr. Yang Jun, engineer of the Research Institute for Geology and Exploration, First Geological Prospecting Bureau of MMI. We also are indebted to Mr. Zhang Haowen, assistant engineer of the same unit, for help with sample preparation, drafting, and data processing.

The manuscript has benefited greatly from reviews by Prof. Brian Windley, Prof. Ian Plimer, and Prof. Francis Saupé, all of whom are heartily thanked for their comments.

Summer 1992

R.B. Trumbull
G. Morteani
Z.L. Li
H.S. Bai

Contents

1	**Introduction**	1
1.1	Regional Tectonic Division of China	2
1.2	The Gold Provinces of Northeastern China	5
1.3	Gold in Eastern Hebei Province	13
2	**Geologic Framework of the Sino-Korean Platform**	17
2.1	Major Orogenic Events	17
2.1.1	Archean: Qianxi Orogeny	19
2.1.2	Early to Middle Proterozoic: Fuping and Wutai Orogenies	19
2.1.3	Middle to Late Proterozoic: Zhongtiao and Yangtze Orogenies	20
2.1.4	Paleozoic: Caledonian and Variscan Orogenies	21
2.1.5	Mesozoic: Indosinian and Yanshanian Orogenies	21
2.1.6	Cenozoic: Pacific Margin and the Himalayan Orogeny	22
2.2	The Precambrian Basement	23
2.2.1	Archean Rocks	24
2.2.2	Proterozoic Rocks	36
2.3	Structural Geology	40
2.3.1	Precambrian Structures	40
2.3.2	Regional Faults and Lineaments	41
2.3.3	Structural Features of the Major Fault Zones	42
2.4	Yanshanian Granites	46
2.4.1	Regional Characteristics and Plate Tectonic Setting	46
2.4.2	Yanshanian Magmatism in Eastern China	49
2.4.3	Yanshanian Granites in Eastern Hebei Province	52

3	**Description of Selected Gold Deposits**	57
3.1	Niuxinshan District	58
3.1.1	Host Rock Lithologies	59
3.1.2	Host Rock Structures	63
3.1.3	Gold Mineralization	64
3.1.4	Wall Rock Alteration	69
3.1.5	Age of Mineralization	73
3.1.6	Fluid Inclusions	74
3.1.7	Stable Isotopic Data	77
3.2	Sanjia District	78
3.2.1	Host Rock Lithology	80
3.2.2	Host Rock Structures	82
3.2.3	Gold Mineralization	84
3.2.4	Wall Rock Alteration	89
3.2.5	Age of Mineralization	91
3.2.6	Fluid Inclusions	93
3.2.7	Stable Isotopic Data	94
3.3	Yuerya District	96
3.3.1	Host Rock Lithology	96
3.3.2	Host Rock Structures	98
3.3.3	Gold Mineralization	99
3.3.4	Wall Rock Alteration	101
3.3.5	Age of Mineralization	102
3.3.6	Fluid Inclusions	102
3.4	Jinchangyu District	103
3.4.1	Host Rock Lithology	104
3.4.2	Host Rock Structures	106
3.4.3	Gold Mineralization	107
3.4.4	Wall Rock Alteration	109
3.4.5	Age of Mineralization	109
3.4.6	Fluid Inclusions	110
3.5	Banbishan District	111
3.5.1	Host Rock Lithology	111
3.5.2	Host Rock Structures	114
3.5.3	Gold Mineralization	115
3.5.4	Wall Rock Alteration	116
3.5.5	Age of Mineralization	117

4	**Aspects of Metallogenesis** 119
4.1	The Age of Mineralization 119
4.1.1	Field Relations 119
4.1.2	Isotopic Age Dates 120

4.2	The Source of Gold 121
4.2.1	Gold in the Qianxi Group: The "Source Bed" Concept 121
4.2.2	Gold in Yanshanian Granites 126
4.2.3	Isotopic Contraints 127
4.2.4	The Mantle Connection 132
4.2.5	Conclusions 133

4.3	The Role of Iron-Rich Host Rocks 134
4.3.1	Banded Iron Formations in Northeastern China 135
4.3.2	Petrography and Composition of Magnetite Quartzites 137
4.3.3	Sulfidization and Gold Mineralization 139
4.3.4	Conclusions 142

4.4	The Role of Yanshanian Granites 142
4.4.1	Granite Compositions 143
4.4.2	Heat Production 146
4.4.3	Conclusions 147

4.5	Structural Controls 148
4.5.1	Regional Structures 148
4.5.2	Local Structures 149
4.5.3	Regional Tectonic Interpretation 150

4.6	Fluid Composition and P-T Conditions of Mineralization 152
4.6.1	Evidence from Fluid Inclusions 152
4.6.2	Stable Isotopic Data 155
4.6.3	Gold Transport and Precipitation 159

5	**Towards a Metallogenetic Model** 163
5.1	Previous Models 168
5.2	The Preferred Model for Eastern Hebei Province 169
5.3	Open Questions 172

XI

| 6 | Comparison with Other Archean-Hosted Gold Provinces | 175 |

Appendix 1 178

Appendix 2 180

References 187

Subject Index 199

1 Introduction

Many of the world's most important gold fields occur in Archean and Early Proterozoic host rocks. The most common association of gold in this setting is with the metamorphosed supracrustal volcano-sedimentary series of the greenstone belts. Prominent examples of the greenstone-gold association are known from nearly all of the Archean cratons, including those in South Africa (Barberton Belt), Zimbabwe (Sebakwian and Bulawayan Belts), Ghana (Birrimian Belt), Western Australia (Norseman-Wiluna Belt), Canada (Abitibi Belt), India (Kolar Schist Belt), and Brazil (Sao Francisco Craton). Reviews of gold geology by Foster (1991), Keays et al. (1989), Hutchinson and Vokes (1987), Foster (1984), and Boyle (1979) provide extensive documentation of this type of gold deposit in the above-mentioned areas.

Conspicuously missing in the international literature are descriptions of the important gold deposits in the Archean craton of northeastern China (with a few exceptions, such as Sang and Ho 1987). According to sources cited by Sang and Ho (1987), China was the world's fourth largest gold producer, with over 3000 deposits. The U.S. Bureau of Mines' estimate of gold production in China in 1989 was 80 tons (U.S. Bureau of Mines 1990). The agency responsible for the operation of gold mines in China depends on their size. The largest mines are operated at the state government level by the Ministry of Metallurgical Industry. Medium-sized and small mines are operated by units of the local county governments.

According to Sang and Ho (1987) and Zhu (1989), the majority of gold production in China comes from deposits in Precambrian uplifts of northeastern China (Shandong, Hebei, Liaoning, and Jilin provinces). The Archean and Early Proterozoic supracrustal rocks in this area, and the gold deposits they contain, are similar in many respects to those of the Archean greenstone belts in Canada, Western Australia, and southern Africa. However, the past decade of research, mostly by Chinese scientists, has shown that there are important differences between the Archean-hosted gold deposits in China and the greenstone belt gold deposits found in other continents.

The need for a metallogenetic model specific to the deposits in northeastern China was recognized by the Ministry of Metallurgical Industry, and this was the main goal of the research on which this text is based. Our investigations were centered around the following six questions considered to be critical for understanding the genesis of these deposits:

1. The age of mineralization.
2. The source of the gold.
3. The role of metamorphic iron formations is gold metallogenesis.
4. The role of granitic intrusions in gold metallogenesis.
5. The structural control of the deposits.
6. The physicochemical conditions of mineralization and the nature of the ore-forming fluids.

This book begins with a brief description of the gold metallogenetic provinces in northeastern China and an introduction to the gold districts of Eastern Hebei province. Chapter 2 summarizes the regional geology of eastern Hebei province with emphasis on the tectonic setting and on the nature of the Precambrian basement. The third chapter then discusses the geological, mineralogical and geochemical nature of five gold districts which were chosen to represent the range of Archean-hosted gold deposit types in this region. Chapter 4 discusses the controls of gold metallogeny based on these examples, and Chapter 5 formulates a tentative metallogenetic model for the deposits. In the final chapter, the Chinese gold deposits are compared with the Archean greenstone-gold deposits of other continents.

1.1 Regional Tectonic Division of China

Before discussing the major gold mining provinces of northeastern China, it is useful to introduce some aspects of the regional geology. The tectonic division of China by Huang (1945) distinguished between stable "platforms" and intervening and/or marginal "mobile" (or fold) belts, and his main tectonic units are, with some modification, still used today. The most recent tectonic compilations are the *Tectonic Map of China* at a scale of 1:4 000 000 (Chinese Academy of Geological Sciences 1979) and the books by Yang et al. (1986), Ren et al. (1987), and Chen (1989). The *Tectonic Map of Asia* at a scale of 1:8 000 000 (Li et al. 1982) combines the tectonic features of China with the rest of Eurasia east of the Mediterranean, and presents a plate tectonic interpretation. Figure 1.1 shows a simplified map of the main tectonic units of China based on Ren et al. (1987), where the platforms and "fold systems" of various ages are clearly shown. Each "fold system" on the figure is made up of one or more "fold belts", and the reader is referred to Ren et al. (1987) and references therein for more complete information.

The tectonic unit which contains the gold districts of eastern Hebei province is the Sino-Korean Platform and the following sections on orogenies, structure, and magmatism focus on this unit. As shown in the following section, other important gold provinces in northeastern China occur in basement uplifts within the marginal fold belts north of the Sino-Korean Platform (e.g., subunits of the Inner Mongolian-Great Hinggan and Jilin-Heilongjiang fold systems in Fig. 1.1).

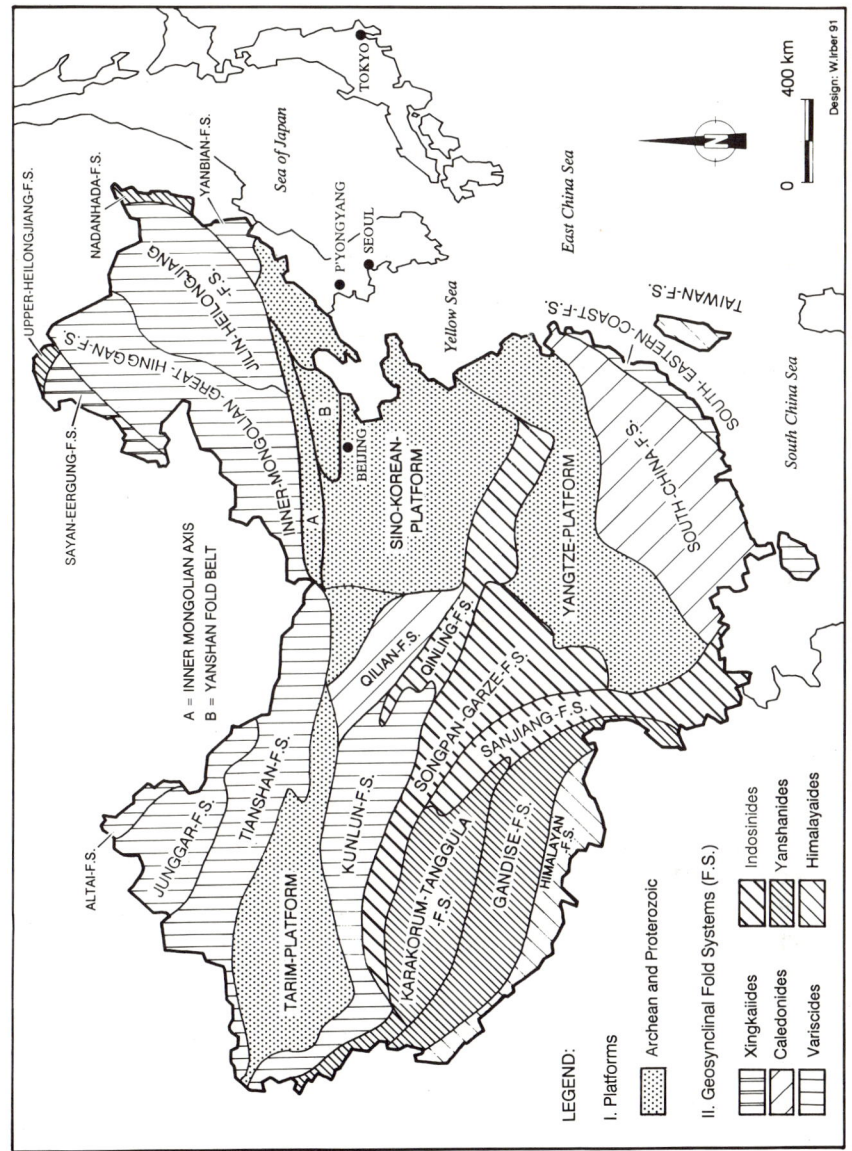

Fig. 1.1. Tectonic map of China. (After Ren et al. 1987)

The Sino-Korean Platform is characterized by several massifs of Early Precambrian metamorphic basement rocks which range from Archean to Late Proterozoic in age, and continental or marine sedimentary cover rocks of Late Proterozoic to Cenozoic age. The Early Precambrian basement massifs (see Fig. 2.1) may represent individual continental "nuclei" which were brought together by later orogenies (Yang et al. 1986). The major part

3

of the platform was consolidated before the end of the Middle Proterozoic, as shown by the extent of Late Proterozoic and younger sedimentary rocks of platformal type (Ren et al. 1987).

The Sino-Korean Platform has not been completely stable since its consolidation in the Proterozoic, and for this reason the term "paraplatform" is preferred by many authors (e.g., Yang et al. 1986; Ren et al. 1987). The influence of Paleozoic (Caledonian and Variscan), Mesozoic (Indosinian and Yanshanian), and Cenozoic (Pacific margin) orogenies is strong in the marginal parts of the platform (Ren et al. 1987). Tectonic reactivation of the platform takes the form of so-called platformal fold belts, deep-seated fracture systems, and zones of magmatism. Two such reactivated units, the so-called Inner Mongolian Axis and Yanshan Fold Belt are singled out in Fig. 1.1A and B, respectively because they are the hosts of the gold deposits discussed in this text.

The strongest orogenic disturbance of the Sino-Korean Platform occurred in the Late Mesozoic (Yanshanian Orogeny), when extensive magmatic activity and NE-trending structures formed as a result of plate interactions on the Pacific margin (Takahashi 1983). It will be shown below that this activity was important for gold metallogenesis in eastern Hebei province and adjacent areas. Cenozoic tectonism on the eastern margin of the Sino-Korean platform was characterized by extension, subsidence, and basin development with associated fissure eruption of alkaline basalts. At this time, the marginal seas (Bohai, Yellow Sea, Sea of Japan) formed by subsidence and extension (Li et al. 1982; Ren et al. 1987). The eastern margin of the platform is still seismically active, and some of the most catastrophic earthquakes in human history have taken place in northeastern China (e.g., Tangshan in 1976).

An important emphasis of recent work by Chinese and foreign geologists has been the application of plate tectonics concepts to interpret the tectonic history of China (Uyeda and Miyashiro 1974; Dickinson 1979; Li et al. 1980, 1982; McElhinny et al. 1981; Klimetz 1983; Terman 1984; Zhang et al. 1984; Maruyama et al. 1989; Wiley et al. 1990). Most studies have focused on the development of the Himalayan orogen (references in Windley 1984) and on the development of the Pacific margin tectonic belt, including the offshore islands and island-arc systems (Uyeda and Miyashiro 1974; Dickinson 1979; Takahashi 1983; Maruyama et al. 1989; Wiley et al. 1990). In older rocks and in regions of interior China far from the continental margins, paleomagnetic evidence and the geologic recognition of paleo-sutures of various ages provide important evidence of plate motions (Zhang et al. 1984). The interpretation of the tectonics of China is still evolving and, although tentative models of the Phanerozoic assembly of eastern Asia have been made (see Li et al. 1982; Klimetz 1983; Zhang et al. 1984; Maruyama et al. 1989; Wiley et al. 1990), these are still highly speculative and not dealt with here in detail. Some aspects of plate tectonics in relation to the Yanshanian Orogeny (Jurassic-Cretaceous) are discussed in Chapter 2.4.

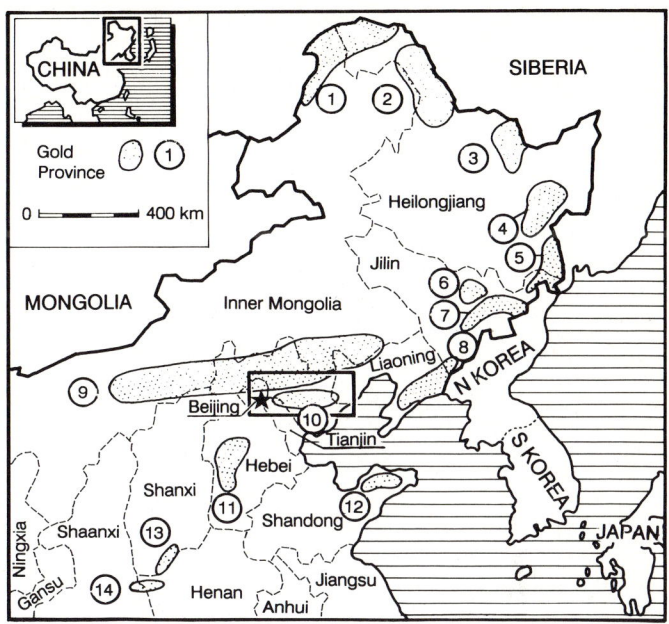

Fig. 1.2. Map of northeastern China showing the distribution of gold provinces discussed in the text. The area of eastern Hebei province covered in this text is *boxed*

1.2 The Gold Provinces of Northeastern China

Important sources of information on the mineral deposits of China in general are Ikonnikov (1975) and the 1:4 000 000 *Metallogenic Map of Endogenic Ore Deposits of China* (Guo 1987). A brief description of some of the most important gold deposits in northeastern China was given by Sang and Ho (1987). Other important sources of information on Chinese gold deposits are the proceedings volumes of the International Conference on Gold Geology and Exploration in Shenyang (Guan and Zhu 1989), and a recent series of monographs entitled *Contributions to the Project of Regional Metallogenetic Conditions of Main Gold Deposit Types in China* edited by the Shenyang Institute of Geology and Mineral Resources (Shenyang 1988a,b, 1989a-d).

Figure 1.2 shows the distribution of 14 gold metallogenetic provinces in northeastern China which were distinguished by Zhang (1979) based on the distribution (i.e., clustering) of deposits, regional tectonic features, geologic environment, and mineralization types. A brief summary of the geologic setting and the mineralization types present in each province is given below. The numbers given in parentheses correspond to the numbers in Fig. 1.2. Where no reference is given, the information cited was obtained from unpublished reports and personal communications of geologists from the Ministry of Metallurgical Industry.

Sang and Ho (1987) discussed five gold metallogenetic provinces in northeastern China. The correlation of these five provinces with the ones delineated below is as follows:

Zhaoye (Sang and Ho) = Ludong (12)
Jiapigou (Sang and Ho) = Longgang-Mudanling (7)
Eastern Hebei (Sang and Ho) = Yinshan-Nuiu'erhushan *and* Yanshan
(9 and 10)
Little Qin Hill (Sang and Ho) = Xiaoqinling (14)
Southern Liaoning (Sang and Ho) = Yingkou-Kuandian (8).

E'erguna Gold Province (1)
The E'erguna gold province is located at the northern border of China in an area of Caledonian and Variscan fold belts (the Heilongjiang and E'erguna fold belts). A metamorphic basement contains rocks of mainly Early Cambrian to Middle- and Late Proterozoic age. Variscan granitic intrusions cover more than half of the exposed area. Jurassic volcanic rocks are found in graben structures with NE–SW and E–W strike. Granite of Yanshanian age (Late Mesozoic) intruded in minor amounts along NE–SW-trending regional fault zones. Auriferous quartz-sulfide veins are found at the intersections of NW–SE and NE–SW-trending faults and around the Yanshanian granitic intrusions. Paleoplacer deposits, mainly in Jurassic sandy conglomerates, are also important gold sources, and more than ten deposits of placer gold are known. Further information on the deposits in this province can be found in Shenyang (1988a).

Xiaoxing-Anling Gold Province (2)
The Xiaoxing-Anling gold province lies in the northern part of the Variscan Daxing-Anling fold belt. The basement rocks consist of Proterozoic schists and gneisses. These are covered by Paleozoic phyllites, immature sandstones, shales, and quartzites, and by Mesozoic andesite-rhyolites. A large Caledonian granite intrusion occurs in the northwestern part of the province, whereas Variscan granites dominate in the eastern and northern parts. Smaller granite intrusions of Yanshanian age are scattered throughout the province. NNE- and NE-trending fault zones influence the distribution of both the Variscan and Yanshanian granite intrusions and of the volcanic rocks. Gold-bearing quartz-sulfide veins constitute the most prominent type of deposit in the Precambrian host rocks, but in the younger volcanic and plutonic rocks epithermal mineralizations, skarn-type deposits, and porphyry-type deposits are also present. Most of the primary gold deposits are associated with the contact zones of the granites; however, the main gold production from the province is from placer gold deposits, with about ten producing mines.

Jiayin-Luobei Gold Province (3)
The Jiayin-Luobei province is located in the northern part of the Jiamosi uplift in a Variscan fold belt north of the Sino-Korean Platform. In the western part of the province widespread Variscan granites and minor occurrences of Proterozoic and Paleozoic sedimentary rocks are found in the Qingheshan Uplift. The central part of the province is formed by a Mesozoic fault-bounded basin (Wulaga Basin). In the eastern part of the province, large areas of Proterozoic schists and gneisses are exposed in the Taipingguo Anticline. Mesozoic rocks occur on both flanks of the anticline. The dominant intrusive rocks are Yanshanian granites, although granites of Variscan age are common in the western part. During the Late Cretaceous, dacites and rhyodacites intruded at a subvolcanic level. These subvolcanic rocks locally host a porphyry-type gold mineralization. Both the plutonic bodies and the subvolcanic intrusions are concentrated along the faults which border the Wulaga Basin. The oldest faults strike NW–SE, and were active intermittently over a long time. E–W- and N–S-trending oblique reverse faults control the distribution of the Mesozoic volcanic rocks. NE–SW- and E–W-trending oblique reverse faults gave passage to the Yanshanian granitic intrusions.
The main gold deposits in the Jiayin-Luobei province are porphyry-type deposits, stratiform metamorphic deposits, and Quaternary placer deposits. The first type is represented by the Tuanjiegou deposit (Li and Liu 1986). The Dongfengshan deposit is a typical stratiform metamorphic gold deposit (Liu 1987; Hao 1989). Other deposits in this province are described in Zhu (1979) and Shenyang (1988a).

Huanan-Laoyeling Gold Province (4)
The Huanan-Laoyeling province includes a fault-bounded Mesozoic sedimentary basin (Wokenhe Basin) in the southern part and Proterozoic basement rocks in the northern part. The latter consist of anatectic granites and metamorphic rocks of Early Proterozoic age exposed in a complex NW–SE-trending anticlinorium. The Wokenhe Basin in the southern part of the province contains a Jurassic carbonaceous clastic sequence and Cretaceous volcanic rocks. The main magmatic rocks are migmatitic granite and granitic gneisses of Early Proterozoic age and Yanshanian granitic intrusions. The regional faults in the Huanan-Laoyeling gold province strike NW–SE and NE–SW. These faults are important, since they form the border of the Mesozoic basin and control in part the distribution of Yanshanian granites and mineralization zones.
Gold mineralization occurs as hydrothermal quartz-sulfide veins in and around the Yanshanian granitic plutons and in the Early Jurassic sandstone-conglomerate series. Placer deposits are also important, and in fact the best-known gold deposit in the province is the placer deposit of the Huanan River. Further information on the deposits in this province can be found in Zhu (1979) and Shenyang (1988a).

Taipingling Gold Province (5)
The Taipingling province is situated in the Yanbian fold belt of Variscan age. In the northern section, schists and gneisses are exposed but the province in general consists of Carboniferous and Permian clastic and volcanic rocks and of Mesozoic volcanics. The most common intrusive rocks are of Variscan age, and they cover a wide area. Granites of Early Proterozoic age occur in the northern section, and Yanshanian granites form only minor bodies which extend along NE–SW-trending fractures.

The most common type of gold deposit in this province consists of hydrothermal quartz-sulfide veins centered about small Variscan granite intrusions. Also mined are epithermal gold-silver deposits hosted by volcanic rocks, and Quaternary placer gold deposits. The primary gold deposits are clearly associated with fracture zones. The gold deposits in the northern section are spatially related to the Variscan granite bodies, and they occur along NE–SW- and NNE–SSW-trending fracture zones. In the central part of the province, gold mineralization is concentrated in N–S-trending fracture zones and is mainly hosted by Permian volcanic rocks and Carboniferous sedimentary rocks. In the southwestern part of the province, gold mineralization is located near the marginal faults of Mesozoic basins. These faults trend mainly E–W and NW–SE. Further information on the deposits in this province can be found in Zhu (1979) and Shenyang (1988a).

Hadaling Gold Province (6)
This province is situated along the northern part of the Juifa River at the northern border of the Sino-Korean Platform in a Variscan fold belt. The geologic units consist of Paleozoic sedimentary and minor volcanic rocks which were folded and overturned at the end of the Permian. Magmatism is represented by granitic intrusions of Variscan and Yanshanian age. The intrusions occur along NE–SW- and NW–SE-striking faults, and they influence the distribution of gold deposits.

Two main areas or subprovinces of gold mineralization occur: in the Erdaodianzi subprovince the dominant ore-controlling structure is an anticlinal fold belt with mainly E–W-trending axes. Gold occurs in quartz saddle-reef veins in the fold hinge zones. The most common host rocks of the gold-bearing veins are Permian carbonaceous shales and hornfelses in the vicinity of a Variscan granite intrusion. Placer gold deposits of Quaternary age are of some importance in this area. The Yongji-Panshi subprovince is built up by Late Carboniferous metavolcanic rocks and Permian carbonaceous sedimentary rocks. Jurassic sedimentary rocks are found in basins bordered by NE–SW- and NW–SE-trending faults. Gold occurs in quartz veins within the Carboniferous metavolcanics. Tang (1986) gives further information on the deposits in this province.

Longgang-Mudanling Gold Province (7)
This gold province is located at the northern edge of the Sino-Korean Platform in the northern part of the Jiaoliao Uplift. The Early Precambrian basement includes amphibolites, plagioclase-hornblende gneisses, greenschists and magnetite-rich quartzites (metamorphic iron formation). Carboniferous and Permian sandstones and shales and Mesozoic volcanic rocks occur locally. Large granite bodies of Variscan age intruded along NW–SE faults mainly in the northeastern part of the province but they are also found as smaller stocks throughout the province. Yanshanian granite bodies are irregularly distributed throughout the province.

Gold mineralization is mainly associated with NE–SW- and NW–SE-trending regional faults. Gold-quartz veins in the Precambrian metamorphic rocks are the dominant type of mineralization. A typical example of this type is the Jiapigou gold deposit (Hu 1989; Nesbitt 1991). Quaternary placer gold deposits are also mined. More information on these deposits is given by S.Q. Wu (1985) and Sang and Ho (1987).

Yingkou-Kuandian Gold Province (8)
The Yingkou-Kuandian gold province is situated in the Jiaoliao Uplift in southern Liaoning province. Within the Jiaoliao Uplift, a Precambrian basement is exposed which consists mainly of high-grade metamorphic rocks of the Archean Anshan Group and Early Proterozoic greenschist-facies metasedimentary rocks of the Liaohe Group. Granite bodies of Middle Proterozoic age cover wide areas. Yanshanian granitic intrusions are also common in a zone of NE–SW elongation.

The gold mineralization is best developed in the Archean and Proterozoic metamorphic rocks. The main types of deposits are gold-quartz veins, pervasively altered fracture zones, metamorphic stratiform deposits and skarn-type mineralizations at the contacts with Yanshanian granites. The gold deposits are mainly associated with E–W-, NNE–SSW-, and NE–SW-trending structures and related secondary fracture and shear zones. Intersections of major fracture zones are particularly important. Placer gold deposits of Quaternary age are of minor economic importance. For more information on the deposits in this province see Sang and Ho (1987) and Shenyang (1988b).

Yinshan-Nuiu'erhushan Gold Province (9)
The Yinshan-Nuiu'erhushan province is a large gold province which extends along the northern border of the Sino-Korean Platform. Most of the province is located within a zone of uplifts known as the Inner Mongolian Axis (Fig. 1.1A) in which Early Precambrian basement is exposed. The basement includes Late Archean rocks which consist of plagioclase-hornblende gneisses, amphibolites, gneisses, schists, marbles, and metamorphic iron formation. Early Proterozoic metamorphic rocks crop out in the western part of the province, and Middle- to Late Proterozoic platformal sedimen-

tary rocks are found in isolated occurrences in the eastern part. Neritic and littoral marine, and intercalated terrestrial sedimentary rocks of Paleozoic age occur on both flanks of the uplift. Mesozoic basins bordered by deep-seated faults contain coal-bearing sedimentary sequences with important volcanic components. Igneous rocks of several ages are present. Middle- and Late Proterozoic intrusions and volcanic rocks span the compositional range from ultrabasic through acid to alkalic. The Proterozoic intrusions are found mainly in the western part of the province. Late Paleozoic and Mesozoic (Variscan and Yanshanian ages), intermediate to acid intrusives are widely distributed. Yanshanian intrusive and extrusive rocks are particularly abundant in the central and eastern parts, and the Yanshanian intrusions are closely related to gold mineralizations.

The main types of gold deposits in the Yinshan-Nuiu'erhushan province are gold-quartz veins and mineralized shear zones in the Archean metamorphic rocks. A spatial association of the deposits with Yanshanian granite intrusions is common. Some vein-type deposits are found within, or in the contact zones of Variscan, and especially, Yanshanian intrusions or within Jurassic volcanic rocks. In addition, mesothermal and epithermal volcanic-related deposits, skarn-type deposits, and Quaternary placer gold deposits are known. The mineralized structures strike E–W, NE–SW, and NNE–SSW. Intersection zones are of major importance for mineralization. Ore bodies are mainly found in secondary fractures related to the regional faults. Important primary gold deposits in the Yinshan-Nuiu'erhushan province are the Jinchanggouliang, Honghuagou, Dashuiqing, and Xiaoyingpan deposits. One of the famous placer gold deposits is the Jinpen deposit, situated in a Mesozoic-Cenozoic basin in the western part of the province. Further information on the deposits in this province can be found in Xia (1986), Guan et al. (1989), and Trumbull et al. (1990).

Yanshan Gold Province (10)
The Yanshan gold province includes the deposits of eastern Hebei province which are the main subject of this book. The gold province is located within the Sino-Korean Platform in the so-called Yanshan platformal fold belt (Fig. 1.1B), which exposes an Early Precambrian basement. The geologic units are dominated by high-grade metamorphic rocks of Archean and Early Proterozoic age. These Early Precambrian rocks are exposed in anticlinoria surrounded by Middle- and Late Proterozoic low-grade metasedimentary rocks of platformal type. Jurassic rocks are distributed in rare outcrops in the northeastern part of the province and they represent a suite of continental and volcanoclastic basin sediments. The earliest magmatic activity recognizable in the Yanshan gold province is represented by Precambrian ultramafic, mafic, and intermediate intrusions associated with E–W-trending structures, mainly in the eastern and western parts of the province. Many of these rocks have been metamorphosed, but textures and field relations clearly reveal their igneous character. A second phase of magmatism took

place during the Mesozoic, and the peak magmatic activity was during the Yanshanian Orogeny. Intrusive rocks of this age are mostly of granitic composition. They form stocks of various sizes, often occurring along fracture zones trending NE–SW and NNE–SSW. Igneous dikes of felsic to mafic composition also formed during the Late Yanshanian Orogeny.

Gold mineralization occurs in gold-quartz veins and pervasively altered fracture zones which are controlled by secondary fractures related to regional-scale NE–SW- and NNE–SSW-trending fault systems. Most gold deposits are spatially associated with granitic intrusions and dikes of Yanshanian age. The most common host rocks are Archean and Early Proterozoic metamorphic rocks. Less important host rocks are the Yanshanian granites. The largest primary gold deposits in the Yanshan province are the Jinchangyu and Yuerya deposits, which are discussed in detail in Chapter 3 of this book. Quaternary placer gold deposits are also worked locally. Further information on the deposits in this province can be found in Sang and Ho (1987), Li (1988), and Shenyang (1989a).

Wutai-Taihang Gold Province (11)
The Wutai-Taihang gold province is located in the interior of the Sino-Korean Platform on the east side of the Shanxi Uplift. The basement rocks include Early Archean high-grade metamorphic sequences overlain unconformably by Middle- and Late Proterozoic low-grade metasedimentary rocks. Cambrian and Ordovician strata occur outside the basement uplift. Magmatism is of only minor importance in the Wutai-Taihang province. As in most areas in the eastern part of the Sino-Korean platform, the magmatic rocks belong to two age groups, Middle Proterozoic and Mesozoic (Yanshanian). The Proterozoic granites are known only from scattered occurrences in the Wutai mountains. The Yanshanian rocks are concentrated in the southeastern and northeastern parts of the province.

Gold mineralization is closely related to Yanshanian felsic dikes which occur along NNW–SSE-trending fault zones. The main types of gold mineralization are syn-metamorphic and post-metamorphic hydrothermal gold-quartz veins in Archean host rocks, and Quaternary placer deposits. The most important deposits of the Wutai-Taihang province are the Yixinzhai and the Shihu deposits. Zhang (1986) describes one of the important deposits (Gengzhuang) in this province.

Ludong Gold Province (12)
The Ludong gold province is the most important in northeastern China in terms of gold production and reserves. It is situated in the eastern Sino-Korean Platform within the so-called Jialiao Uplift. The basement rocks in the province are Late Archean and Early Proterozoic high-grade metamorphic rocks. Cretaceous sandstones and shales, conglomerates, and marls occur within extensional basins, and at the basin margins Late Mesozoic and Tertiary felsic volcanic rocks are found. The magmatic rocks in the Ludong

province are dominated by granitic intrusions of Yanshanian age. The main structures in the province are E–W-trending folds in the basement complex and large-scale NE–SW- and NNE–SSW-trending fault zones of Mesozoic age. The western part of the gold province is bounded by the major NE–SW-trending Tancheng-Luliang fault.

The gold mineralization is mainly associated with NE–SW-trending Mesozoic faults and related secondary fractures in the granites. The mineralization types are gold-quartz veins and altered fracture zones. The main primary gold deposits in the province are the Linglong and Jiaojia deposits. Some placer gold deposits are also known. Descriptions of the deposits in this province can be found in Huang (1986), Sang and Ho (1987), Shenyang (1989c), Lu et al. (1989), and Zhou and Fan (1989).

Zhongtiaoshan Gold Province (13)

The Zhongtiaoshan province is located in the northern part of the western Henan uplift at the southern margin of the Sino-Korean Platform. The basement consists of Late Archean metamorphic rocks including mica schists, amphibolites, and hornblende gneisses; and Early to Middle Proterozoic quartzites, conglomerates, and marbles with minor mica-schists and phyllites. The main structures of the province trend NNE–SSW. The basement structures include tight folds overturned to the west. Faults are mainly NNE-directed overthrusts and E–W-trending fracture zones. Magmatism is dominated by Yanshanian granitic intrusions and associated felsic dikes concentrated in a zone along the northern part of the province.

Gold-quartz veins and auriferous porphyry copper deposits are the main types of gold mineralization. Whereas the quartz veins are mostly found in the basement rocks at least 10 km away from Yanshanian intrusions, the porphyry-type deposits formed within or in the contact zone of granodiorites. Gold placers also occur in the province. Further information can be found in Sha (1986).

Xiaoqinling Gold Province (14)

This province lies in the contact zone between the Sino-Korean Platform to the north and the so-called Qinling fold belt to the south. The latter is a complex zone of superimposed Paleozoic and Mesozoic folds situated between the Sino-Korean and the Yangtze Platforms. Basement rocks are exposed in an E–W-trending anticlinorium bounded by major fault zones. The rocks consist of high-grade Late Archean amphibolites, migmatitic gneisses, and metamorphic iron formations; and Early Proterozoic metavolcanic rocks, overlain by nonmetamorphosed Late Proterozoic clastic and carbonate sedimentary rocks. Jurassic and Tertiary sedimentary rocks occur outside the basement uplift. Magmatic rocks are represented by Proterozoic granites situated at the southern margin of the province and by granite intrusions of Yanshanian age within the anticlinorium and along the bordering faults.

Thrust faults are the main mineralized structures and they are mostly developed in the axial zone of a secondary anticlinorium. Less important for the mineralization are N–S-trending oblique normal faults. The gold mineralization is closely related to granites of Yanshanian age. Further information on the deposits in this province can be found in Wang (1987), Sang and Ho (1987), and Shenyang (1989b).

1.3 Gold in Eastern Hebei Province

The gold deposits of eastern Hebei province are included in the so-called Yanshan Gold province described above (number 10 in Fig. 1.2). Yu et al. (1989) reported 197 gold deposits and occurrences in eastern Hebei province. The distribution of the most important of these is shown in Fig. 1.3. Two important features of the distribution of gold deposits are apparent from the map. First, the deposits are mainly found in the Archean and Early Proterozoic basement and second, many deposits show a spatial association with Mesozoic granitic intrusions.

Two main gold deposits types (excluding placers) occur in eastern Hebei province according to the classification of Zhu (1989), namely, metamorphic-hosted and granite-hosted deposits. The metamorphic-hosted type is by far the more common, and most gold deposits occur in Archean mafic metamorphic rocks. Both the metamorphic-hosted and granite-hosted gold deposits are associated with secondary faults and fractures related to major district-wide fault zones and/or lineaments.

The largest gold mining districts in Eastern Hebei province are briefly described below. The letter preceding each district name corresponds to those on Fig. 1.3. For the purpose of location, the nearest village or town is given in the descriptions, although these towns cannot be shown at the scale of Fig. 1.3.

A) Niuxinshan

The Niuxinshan gold mining district is located about 40 km west of the town of Qinglong. The mineralization is developed in quartz-sulfide veins surrounding a Yanshanian granite intrusion. The host rocks are Early Archean amphibolites (Qianxi Group), and to a lesser extent, Yanshanian granite. Two deposits occur in the district, the Niuxinshan and Huajian deposits. A detailed description of the Niuxinshan district is given in Chapter 3.1.

B) Sanjia

The Sanjia gold mining district is located about 20 km north of the town of Qinglong. It has the same geologic character as the Niuxinshan district. Quartz-sulfide veins with gold are found in Early Archean amphibolites (Qianxi Group), and rarely in Yanshanian granite. Three gold deposits

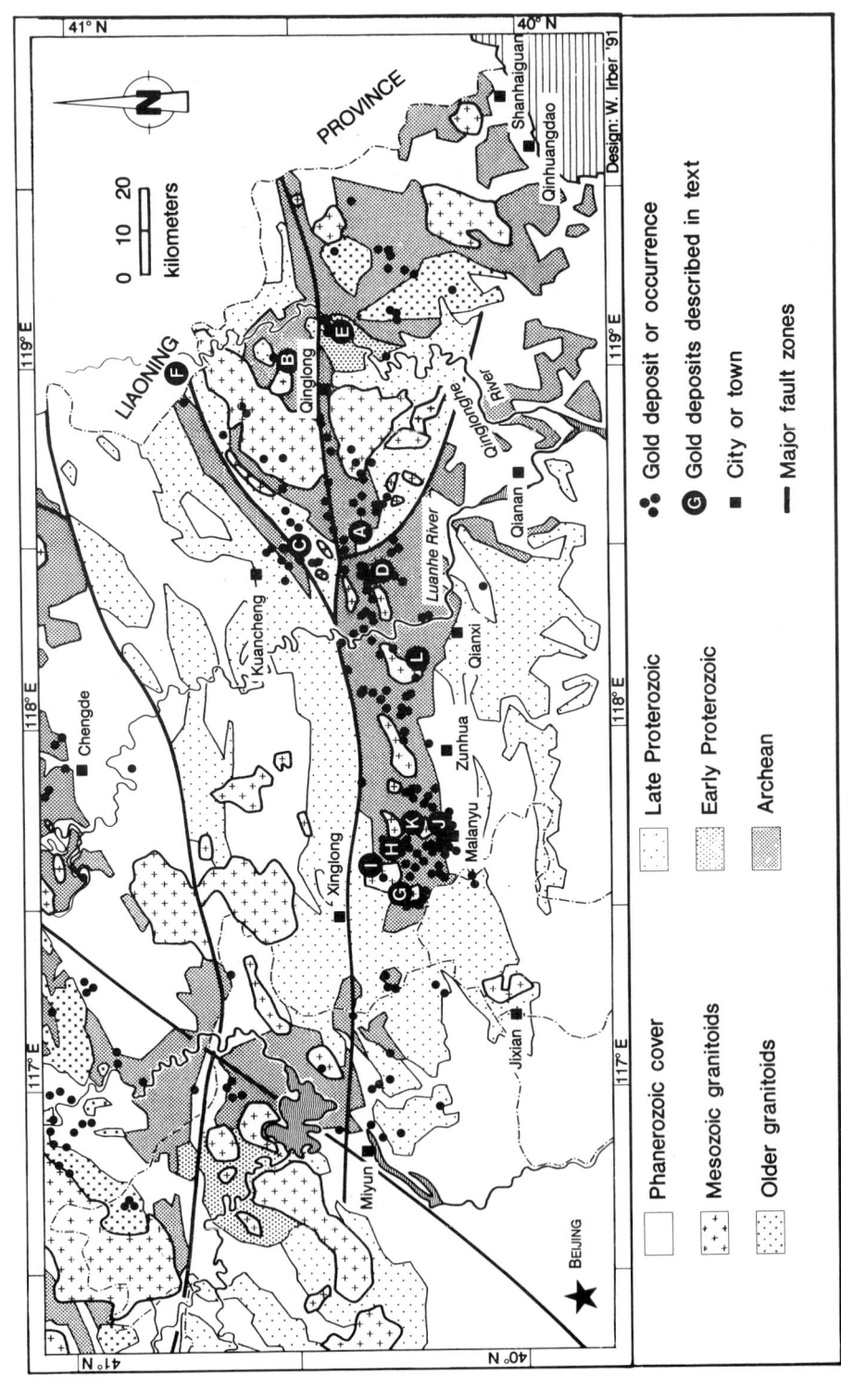

occur within the Sanjia district: Sanjia, Xinglonggou, and Wangtoushan. The deposits are described in detail in Chapter 3.2.

C) Yuerya
The Yuerya gold mining district is located near Yuerya village in Kuancheng county. The mineralization consists of hydrothermal quartz-pyrite veins and vein-parallel disseminated zones near the contact of a Yanshanian granitic intrusion and Late Proterozoic dolomitic limestone (Changcheng system). The main host rock is granite. The Yuerya deposit is described in detail in Chapter 3.3.

D) Jinchangyu
The Jinchangyu district is located southwest of Xiaying village in Qianxi county. The gold mineralization is mainly developed in hydrothermal polymetallic quartz veins in shear zones. The main host rocks are Early Archean amphibolites and plagioclase-hornblende gneisses (Qianxi Group). A detailed description of the district is given in Chapter 3.4.

E) Banbishan
The gold mining district of Banbishan is located near Shuangshanzi village in Qinglong county. The mineralization consists of thin quartz-sulfide stringers and disseminations in shear zones. The host rocks are Early Proterozoic quartz-rich schists and conglomerates (Zhuzhangzi Group). The district includes the Banbishan and Zhangzhangzi deposits, or which only the Banbishan deposit is currently mined. A detailed description of the Banbishan deposit is given in Chapter 3.5.

F) Baizhangzi
The Baizhangzi district is located near Baizhangzi village in Lingyuan county. The mineralization consists of hydrothermal quartz veins related to a Yanshanian granite intrusion. The ore veins are hosted partly by Yanshanian granite and partly by Middle Proterozoic sandstone (Changcheng system).

G) Madi
The Madi district is located in the Paomaochang commune in Xinglong county. The mineralization occurs in quartz veins hosted by granoblastic gneisses and amphibolites of the Archean Qianxi Group.

◄ **Fig. 1.3.** Simplified geologic map of eastern Hebei province with gold deposits and occurrences marked by *black dots*. The deposits described in the text are: *A*, Niuxinshan; *B*, Sanjia; *C*, Yuerya; *D*, Jinchangyu; *E*, Banbishan; *F*, Baizhangzi; *G*, Madi; *H*, Huashi; *I*, Daoliushui; *J*, Malanyu; *K*, Maoshan; *L*, Majiayu

H) Huashi
The Huashi gold mining district is located 4 km west of the Sibazi commune in Xinglong county. The mineralization occurs in quartz veins within faults developed in Early Archean granoblastic gneisses.

I) Daoliushui
The Daoliushui gold mining district is located near Jinshan commune in Xinglong county. Gold mineralization occurs in hydrothermal quartz-sulfide veins hosted by Yanshanian granites and by Archean amphibolites and granulites (Qianxi Group).

J) Malanyu
The Malanyu gold district is located near the village of Malanyu about 20 km northwest of Zunhua. The gold is worked from several Quaternary river terrace sediments. The main gold enrichment is found in sandy and conglomeratic sediments. Underlying the sediments are Archean metamorphic basement rocks and Yanshanian granodioritic intrusions.

K) Maoshan
The Maoshan gold deposit is located near the village of Maoshan in Zunhua county. The gold mineralization occurs in quartz-sulfide veins in Archean metamorphic rocks near the contact of the Yanshanian Maoshan Granite. Some ore veins also occur within the granite.

L) Majiayu
The Majiayu district is located near the Gaojiadian commune in Qianxi county. Gold mineralization occurs in quartz-sulfide veins hosted partly by Early Archean amphibolites and leptites of the Qianxi Group, and partly by Middle Proterozoic quartzites and marbles.

2 Geologic Framework of the Sino-Korean Platform

The purpose of this chapter is to summarize the geologic setting of the gold deposits in eastern Hebei province within the context of the Sino-Korean Platform. Following a brief summary of the main orogenic events which affected the Sino-Korean Platform, this discussion concentrates on aspects of the geology which are most relevant to the genesis of the gold deposits, namely:

1. the lithology and metamorphic history of the Precambrian basement,
2. the structural geology of the Precambrian basement and the timing and orientation of major fault zones,
3. the Yanshanian (late Mesozoic) magmatism, which in this area involved dominantly granitic intrusions and dikes.

A complete summary of the geology of the Sino-Korean Platform is beyond the scope of this text and the reader is referred to *The Geology of China* by Yang et al. (1986), *Geotectonic Evolution of China* by Ren et al. (1987), *Tectonics of China* by Chen (1989), the *Atlas of the Paleogeography of China* (H.Z. Wang 1985), and the *Tectonic Map of Asia with Explanatory Notes* (Li et al. 1982) for recent and comprehensive reviews of Chinese geology.

2.1 Major Orogenic Events

Table 2.1 summarizes the main orogenic events recognized in China and their approximate correlations with those of North America and Europe. This table is taken from Yang et al. (1986) and, following these authors, it distinguishes between "tectonic stages" and "orogenic movements". The "movements" are phases of intense folding, faulting, and magmatism that may be equated with the terms "orogeny" or "orogenic event". The tectonic "stages" cover much longer periods of time than the "movements" and include phases of quiescence and sedimentation (Yang et al. 1986). Table 2.1 also shows the most important geologic events which are attributed to the various orogenies. The reader should be aware that these "main geologic events" are interpretations which are by no means universally accepted. For example, Yang et al. (1986) and Ren et al. (1987) attribute the Variscan fold belts in northeastern China to a Late Permian convergence of the Sino-Korean Platform with the Siberian Platform. On the other hand, McElhinny

Table 2.1. Chronological table of the major orogenic events in China. (Yang et al. 1986)

Geologic time scale		Tectonic stages		Orogenic movements	Main geologic events	Orogenic movements of Laurasia		
							Europe	N. America
Cainozoic	Quaternary — 2.0 — Neogene — 24.6 — Eogene — 65 —	Megastage of Pangaea (Gondwana) disintegration	Himalayan Stage	~Himalayan 2~ ~Himalayan 1~ ~Yanshanian 3~	Upheaval of Qinghai Tibet Plateau Collision of Himalaya and Gangdise Opening of South China sea	Young Alpedic	~Rodanian~ ~Savian~ ~Pyrinean~	~Laramidian~
Mesozoic	Cretaceous — 144 — Jurassic — 213 — Triassic — 248 —		Yanshanian Stage	~Yanshanian 2~ ~Yanshanian 1~	Collision of Gangdise and Qiangtang Activation of East China continental margin	Old Alpedic	~Late Cimmerian~ ~Early Cimmerian~	~Nevadian~
			Indonisian Stage	~Indonesian 2~ ~Indonesian 1~	Convergence of Yangzi and N. China to form Palasia		~Pfalzian~	
Late Palcozoic	Permian — 286 — Carboniferous — 360 — Devonian — 408 —	Megastage of Pangaea (Laurasia) formation	Hercynian (Variscan) Stage	~Yiningian[a]~ ~Tianshanian~ ~Qilianan~	Convergence of N. China and Siberian–Mongolia Disruption of W. border of Yangzi Platform Formation of S. China Caledonides and close of Quilian troughs	Hercynian	~Sudetian~ ~Bretonian~ ~Erian~	~Alleghenian~ ~Acadian~
Early Palcozoic	Silurian — 438 — Ordovician — 505 — Cambrian — 600 —		Caledonian Stage	~Gulangian~ ~Xingkaian~ ~Chengjiangian~	Formation of Junggar and other median massifs	Caledonian	~Taconian~ ~Salairian~ ~Assyntian~	
Late Proterozoic	Sinian — 850 — Qingbaikouan — 1050 —	Megastage of platform formation	Jinningian Stage	~Jinningian~	Formation of Yangzi Platform and Qaidam Massif etc.		~Gothian~	~Grenvillian~
Middle Proterozoic	Jixianian — 1400 — Changchengian — 1850 —			Sibaoan ~Zhongyuean~ ~Luliangian~ (Zhongtianoan)	Formation of N. China and Tarim Protoplatforms		~Karelian~	~Hudsonian~
Early Proterozoic	Hutuoan — 2200–2300 — Wutaian — 2500–2600 —		Luliangian Stage	~Wutaian~ ~Fupingian~	Formation of Ordos and Jilu Nuclei		~Belomorian~	~Kenoran~
Archean	Fupingian — 2900–3000 — — 3800 — Hadean — 4500 —	Megastage of continental nuclei formation	Fupingian Stage					

[a] Or Nilkaan.

et al. (1981) presented paleomagnetic evidence which indicated that the Sino-Korean and the Yangtze Platforms were widely separated from the Siberian platform and from each other in the Late Permian.

The following discussion of the major orogenic events in China follows the tectonic interpretations of Yang et al. (1986) and Ren et al. (1987) except where other references are given. We use the term "orogeny" and the subordinate term "event" in the same sense as "movement" in Table 2.1 as explained above. Emphasis is given to the Sino-Korean Platform and adjacent areas of northeastern China. For details of the tectonic development of other areas of China the reader is referred to Ren et al. (1987) and the references given therein.

2.1.1 Archean: Qianxi Orogeny

The earliest orogeny recognized in China is the Early Archean (pre-3 Ga) Qianxi Orogeny (Ma and Wu 1981). The evidence for this is mainly given by Early Archean isotopic ages; the geologic evidence has not survived the deformation and granulite-facies metamorphism of the 2.5 Ga Fuping Orogeny (see below), which affected all known Archean areas in northeastern China.

Little can be said for sure of the nature of the Qianxi Orogeny because of the intense 2.5 Ga overprint. Indeed, Ren et al. (1987) do not recognize a tectonic event at all prior to the Fuping Orogeny, although they acknowledge the presence of rocks older than 3 Ga. Ma and Wu (1981) attribute the granulite-facies regional metamorphism in eastern Hebei province to the Qianxi Orogeny, but Sm-Nd dating by Jahn et al. (1987) and Jahn and Zhang (1984) suggest that the granulite-facies metamorphism, at least in the area north of the Luanhe River (Fig. 1.3), is more likely of Fupingian age (2.5 Ga).

The isotopic ages of 3.4–3.5 Ga reported by Jahn et al. (1987) and Huang et al. (1986) from amphibolite enclaves in orthogneisses south of the Luanhe River are interpreted as the age of mantle-derived basaltic magmatism. Liu et al. (1990) concluded that a sialic basement existed prior to 3.6 Ga based on a $^{207}Pb/^{206}Pb$ zircon age from quartzite. Although it is certain that pre-Fupingian metamorphism and tectonism affected these Early Archean supracrustal rocks, the geologic evidence for such events has apparently been largely destroyed.

2.1.2 Early to Middle Proterozoic: Fuping and Wutai Orogenies

The Fuping Orogeny at 2.4–2.6 Ga involved widespread granulite-facies metamorphism with intense ductile deformation, the intrusion of granitic plutons, and mafic volcanism. Most of the Fuping magmatic rocks which have been studied in eastern Hebei province have near-primitive Sr and Nd isotopic signatures and are interpreted to represent new additions to the continental crust (Pidgeon 1980; Jahn and Zhang 1984; K.Y. Wang et al. 1985). The Fuping Orogeny affected all the known Early Archean rocks of northeastern China, and most isotopic ages from the Archean exposures fall in the "Fupingian" range (Sun and Lu 1985; Yang et al. 1986; Jahn and Ernst 1990). The typical Fuping structural style, according to Ma and Wu (1981), involved the formation of gneiss domes (such as in the Qianan region of eastern Hebei province, see Chap. 2.2.1.1) surrounded by tightly folded belts. The structural trends of the Archean rocks in eastern Hebei province imparted by the Fuping Orogeny are NNE to NE (Ma and Wu 1981).

The Wutai Orogeny (2–2.2 Ga) was a time of major accretion of marginal fold belts and separate continental blocks to form the Sino-Korean Platform (Ren et al. 1987). The Wutai Orogeny is best expressed in fold belts along the margins of the Archean "continental nuclei", where Middle Proterozoic rocks are metamorphosed to lower amphibolite-facies assemblages. In eastern Hebei province, the Wutai Orogeny is represented by a "tectonothermal event" at about 2.2 Ga resulting in amphibolite-facies metamorphism and NE–SW-trending open folds (Sun 1984). In contrast to the ductile deformation of the earlier orogenies, deep-seated fault zones associated with plutonic activity formed during the Wutai Orogeny (Ma and Wu 1981), testifying to crustal rigidity.

2.1.3 Middle to Late Proterozoic: Zhongtiao and Yangtze Orogenies

The Middle Proterozoic (1.7–1.9 Ga) Zhongtiao Orogeny (also named Luliang) caused further consolidation of, and marginal accretion to the Early Precambrian platforms. In northern China a continuous "protocontinent" was formed by the joining of the Sino-Korean and Tarim Platform (see Fig. 1.1). Local extension of the platform in northern China was marked by the intrusion of extensive mafic dike swarms, anorthosite, and rapakivi granite suites and by the formation of sedimentary rift basins (Ma and Wu 1981). One of these structures is the so-called Yanshan platformal fold belt or Yanshan Trough (also referred to in the literature as the Yanshan settling belt, Yanshan geosyncline, and Yanshan aulocogen) in the northern part of eastern Hebei province. This structure is of relevance to this study because it is host to most of the gold districts described in Chapter 3. The location is marked by the letter B in Fig. 1.1. It is an E–W-trending zone marked by thick accumulations of Middle and Late Proterozoic quartzites and flysch-like sedimentary rocks with K-rich mafic and intermediate volcanic rocks in the basal parts of the sequence (Ma and Wu 1981; Sun and Lu 1985).

The Late Proterozoic Yangtze Orogeny mainly affected southern and southeastern China. The Yangtze Orogeny was important in the consolidation of the Yangtze Platform, and it is commonly subdivided into the Jinning (850 Ma) and Chengjiang (700 Ma) Events (Ren et al. 1987). In northern China evidence for the Jinning Event is found along parts of the southern margin of the Tarim and Sino-Korean Platforms; the Chengjiang Event is not expressed.

2.1.4 Paleozoic: Caledonian and Variscan Orogenies

The main Caledonian orogenic events in China took place in the Late Ordovician to Early Devonian periods, although an Early Caledonian "Xingkai Event" of Middle Cambrian age is recognized in west-central China and in the Indochina peninsula (Ren et al. 1987). The most important areas of Caledonian activity are in southern and western China, where extensive accretionary fold belts represent the addition of eugeosynclinal material (island arcs and marine marginal basins) to the Yangtze and Tarim platforms. Along the northern border of China and in Mongolia, Late Caledonian (Silurian) subsidence and folding are recorded, but here the main phase of orogenic activity was Variscan, when the final collision of the Sino-Korean platform with the Siberian platform took place (according to Ren et al. 1987).

The Variscan Orogeny (Devonian to Permian) involved the collision of the "North China continent", i.e., the Tarim and Sino-Korean Platforms, with the Siberian-Mongolian platform. Variscan fold belts are developed extensively in northwestern and northeastern China, but they are also represented on the southern margin of the Yangtze platform in southeastern China. The well-known Tianshan and Altay mountain ranges in western China are considered typical of the Variscan fold belts by Ren et al. (1987). Several phases of Variscan Orogeny are distinguished by Ren et al. (1987), to which the reader is referred for details. In northeastern China the Variscan Orogeny produced, in addition to fold and thrust belts, large amounts of dioritic to granitic intrusions and minor ultramafic intrusions, which tend to form along major fault zones and lineaments. These rocks are concentrated around the northern and southern borders of the Inner Mongolian Axis in the Sino-Korean Platform (marked by A in Fig. 1.1), and farther to the northeast. Variscan volcanic rocks are not abundant, and only minor intermediate lavas and pyroclastics are known in Permian strata. The volcanic rocks include andesites, pyroxene-bearing andesites, crystal and vitric tuffs and andesitic breccias which formed in the so-called Inner Mongolian eugeosyncline (Yang et al. 1986).

2.1.5 Mesozoic: Indosinian and Yanshanian Orogenies

The Indosinian Orogeny (Triassic) had only minor effects within the Sino-Korean Platform, but was important in southwestern and central China (Fig. 1.1). A narrow zone of Triassic folding and magmatism in central China represents convergence of the Sino-Korean and the Yangtze Platforms (Yang et al. 1986). According to these authors, eastern China was consolidated into a single continental mass by the Indosinian Orogeny. Westward subduction of oceanic lithosphere of the Izanagi plate beneath the eastern margin of the Sino-Korean and Yangtze platforms began in the Late

Triassic. In southwestern China and in the adjacent parts of Indochina, the Indosinian Orogeny is represented by extensive fold belts resulting from the northward convergence of Tethyan island arcs with the Asian continent (Ren et al. 1987).

The Yanshanian Orogeny (Jurassic to Cretaceous) produced wide tectono-magmatic belts on the margins of the Chinese platforms in both eastern China and southwestern China due to continued subduction of oceanic lithosphere (Pacific and Tethyan, respectively) which had begun in the Indosinian Orogeny.

In eastern China, the Yanshanian Orogeny produced a fundamental change in the orientation of structures within the continent from dominantly E–W (related to the N–S accretion and collisions of continental blocks) to dominantly NNE–SSW (caused by interaction with the Pacific margin). The orogeny involved extensive volcanism, plutonism, and the development of elongate NNE-trending fault-bounded basins to a distance of over 1000 km inland from the continental margin. Along the southeastern coastal region of China and the island of Taiwan, regional metamorphism of greenschist grade developed, but elsewhere in eastern China the only metamorphic effects of the Yanshanian Orogeny were contact metamorphism related to igneous intrusions (Yang et al. 1986). The Yanshanian was also the period of most intense metallogenetic activity in eastern China, as discussed more fully in Chapter 2.4.

In western and southeastern China the Yanshanian Orogeny developed during the Late Cretaceous with the formation of extensive fold belts and widespread volcanic and plutonic activity, and with regional uplift. The events reflect continued subduction of the Tethyan oceanic crust and island-arc accretion to the Asian continent (Yang et al. 1986; Ren et al. 1987).

The subdivision of the Yanshan Orogeny is still debated. According to most authors, the Yanshan Orogeny spans the time from Early Jurassic to Late Cretaceous, and can be divided into three phases, as shown on Table 2.1 (Yang et al. 1986; Ren et al. 1987). According to this classification, the third and last phase of the Yanshanian Orogeny terminated in the Early Tertiary. However, Wan and Zhu (1989) argued that the regional stress orientation of the orogenic phase beginning in the Late Cretaceous was different from that of the earlier Yanshanian phases. They suggested that the term Yanshanian Orogeny be limited to the Jurassic, and that the Cretaceous events should be called the Sichuan Orogeny. This suggestion has not yet found wide acceptance and is not followed in this text.

2.1.6 Cenozoic: Pacific Margin and the Himalayan Orogeny

The Cenozoic orogenic events involved, on the one hand, continuing subduction-related magmatism and associated crustal extension at the Pacific margin in eastern China, and on the other, the collision of the Indian subcontinent with Eurasia (Himalayan Orogeny), which affected mostly southwestern and central China. The Himalayan Orogeny produced dominantly compressional features and tremendous regional uplift of the Qinghai-Tibet plateau. Within the Asian continent to the north of the Himalayas, reactivation of earlier mountain ranges and major strike-slip faults developed due to the post-collisional indentation of India into Asia (Molnar and Tapponier 1975).

In eastern China, deep Tertiary and Quaternary sedimentary basins developed in response to crustal extension. Tectonic features are dominated by NNE-trending extensional and strike-slip faults. The opening of the Sea of Japan and of the South China Sea due to back-arc spreading took place from the end of the Cretaceous to the Late Tertiary (Uyeda and Miyashiro 1974; Dickinson 1979). The Bohai Sea and the Yellow Sea are Quaternary features caused by subsidence related to crustal thinning and extensional faults. Cenozoic magmatic activity in eastern China is small in scale and is dominated by mafic volcanism.

2.2 The Precambrian Basement

The Sino-Korean platform contains nearly all of the known exposures of Archean and Early Proterozoic rocks in China (Yang et al. 1986). These exposures consist of high-grade metamorphic rocks of igneous and sedimentary origin, and they represent the ancient crystalline basement of the platform. Middle and Late Proterozoic rocks in China are widespread not only in the Sino-Korean Platform but also in the Tarim Platform to the west, and in the Yangtze Platform to the south (Yang et al. 1986). In the other platforms, the Middle Proterozoic constitutes the crystalline basement but in northeastern China the Middle- and Late Proterozoic rocks comprise weakly metamorphosed platform sequences of carbonate rocks and clastic sedimentary rocks. The distribution of Precambrian exposures in northeastern China is shown in Fig. 2.1 based on the *Tectonic Map of the People's Republic of China* at a scale of 1:4 000 000 (Chinese Academy of Geological Sciences 1979). The reader is referred to Yang et al. (1986), Sun and Lu (1985), and Ma and Wu (1981) for general reviews of the Precambrian of China.

It was shown in Chapter 1.3 that, in eastern Hebei province, the Archean rocks host most of the gold deposits (see also Fig. 1.3). Therefore the emphasis of the chapter is on a description of the Archean rocks. The Early Proterozoic rocks in eastern Hebei province host some minor gold deposits and occurrences, for example the Banbishan deposit described in Chapter 3.5. The Late Proterozoic platform rocks are rarely hosts to gold deposits, although a notable exception is the Yuerya gold deposit, which is described in Chapter 3.3.

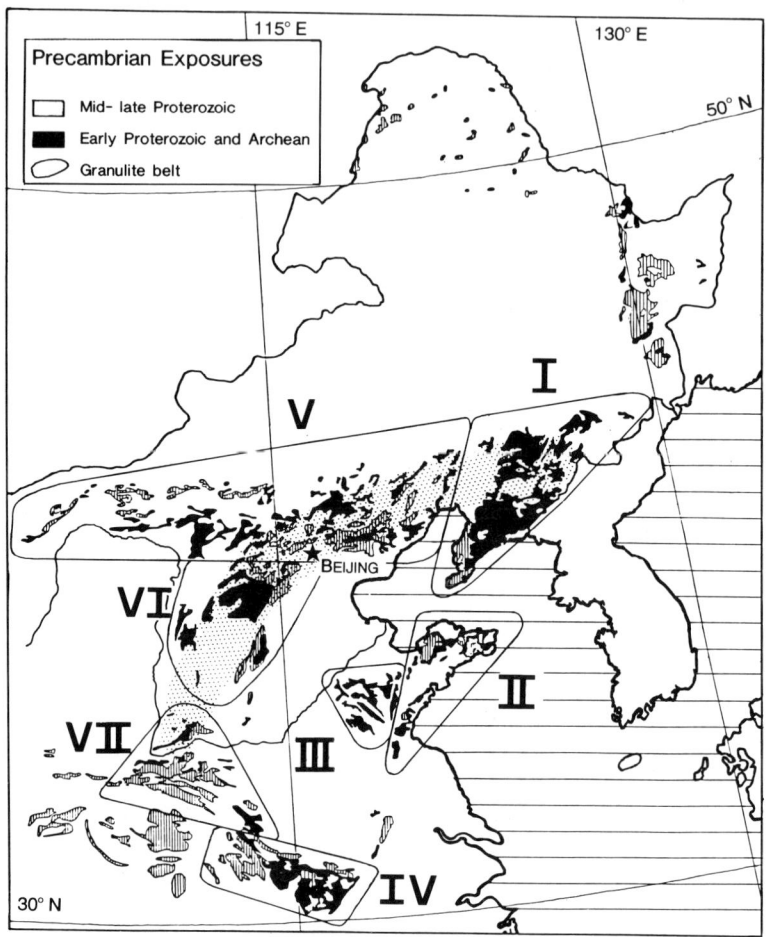

Fig. 2.1. The distribution of Precambrian outcrops in northeastern China, from the Chinese Academy of Geological Sciences (1979) 1:4 000 000 scale tectonic map of China. *Roman numerals* refer to the Precambrian massifs listed in Table 2.2

2.2.1 Archean Rocks

The Precambrian rocks exposed in eastern Hebei province represent nearly 3000 million years of earth history, and the province contains the best-studied Archean rock sequences of the Sino-Korean Platform. The oldest known rocks are Early Archean granulite-facies metasedimentary rocks in the Qianan area from which zircons have recently been dated at 3600 Ma (Liu et al. 1990). A wealth of data on the lithology, mineral composition, whole-rock chemistry, and radiogenic isotopic composition of the Archean rocks has been published, and research on these rocks is very active (Sun and Wu 1981; Zhang and Cong 1982; Jahn and Zhang 1984; Sun 1984; K.Y. Wang et al. 1985, 1990; R.M. Wang et al. 1985; Huang et al. 1986; Jahn et al. 1987; Sills et al. 1987a; Jahn 1990, 1991; Liu et al. 1990). Before the rocks of eastern Hebei province are discussed in detail, it is worth reviewing some general features of the Archean geology in northeastern China together with observations from other areas.

Regional Comparisons

Yang et al. (1986) distinguished seven areas in northeastern China where Archean rocks are exposed, and these are marked in black in Fig. 2.1. In all of these, the Archean is characterized by complex, polymetamorphic sequences of high grade supracrustal rocks, frequently showing evidence of migmatization and intrusion by one or more generations of felsic plutonic rocks now represented by orthogneisses. It is impossible to understand the evolution of these rock complexes without abundant isotopic age data, and these data are only beginning to become available in some areas. For this reason, the correlations among the separate Archean complexes are tentative, and even within a single complex several schemes of nomenclature and subdivision may exist. Table 2.2 lists the correlation and nomenclature of the Archean and Early-Proterozoic lithostratigraphic groups in northeastern China according to Yang et al. (1986).

According to Yang et al. (1986), all the Archean rocks in northeastern China have been metamorphosed to at least the upper amphibolite facies, and most show evidence of anatexis. Retrograde greenschist facies and lower amphibolite facies assemblages are common. Granulite-facies assemblages are found in three of the seven areas, and the distribution of

Table 2.2. Correlation of rock units in seven Early Precambrian exposures in northeastern China. (After Yang et al. 1986)

Region	I	II	III	IV	V	VI	VII
	Liaoning and Jilin	Eastern Shandong	Western Shandong	Huaiyang	Yanshan	Wutai Taihang	Eastern Qingling
Early Proterozoic	Liaohe Group	Fenzishan Group		Hongan Group	Zhuzhangzi Group	Wutai Group	Songshan Group
Archean	Anshan Group	Jiodong Groupd	Taishan Group	Dabie Group	Dantazi Group	Longquanguan Group	Dengfeng Group
					Qianxi Group	Fuping Group	

these led Yang et al. (1986) to propose the concept of a "northern granulite belt" (stippled in Fig. 2.1).

In spite of the uncertainty caused by intense metamorphism, the Archean rocks can be divided into two categories based on their protoliths: supracrustal rocks (metavolcanic and metasedimentary) and plutonic rocks (orthogneisses). Most of the Archean supracrustal rock types present include amphibolites, pelitic (garnet-sillimanite) gneisses, magnesian marbles and calc-silicate gneisses, metamorphic banded iron formations, quartzites, graphitic schists, and pelitic schists. The common Archean gray felsic gneisses, mainly biotite-plagioclase gneisses or hornblende-plagioclase gneisses, are of ambiguous origin. In some cases, field relations suggest that the felsic gneisses, in particular the granoblastic ones, were intercalated with the supracrustal rocks and may represent immature arc sediments or tuffs (Liu et al. 1990). In other cases there is no good evidence of the protolith character.

The plutonic rocks, now represented by orthogneisses, have a wide range of granitic compositions from tonalite to granite. Where the rocks have been studied in detail, for example in Liaoning province (Ernst et al. 1988; Wang et al. 1990) and in eastern Hebei province (Liu et al. 1990), it was found that there were several generations of intrusion. Early intrusions contain enclaves of supracrustal rocks (amphibolites, aluminous gneisses). These early orthogneisses resemble the TTG (trondhjemite-tonalite-granodiorite) gneisses common in Archean terranes throughout the world (Windley 1984). Both the TTG orthogneiss and their enclaves have been metamorphosed at the upper amphibolite to granulite facies. Later intrusions, which formed after the peak of high-grade metamorphism, are of more granitic composition (granodiorite-quartz monzonite-granite).

The relative proportions of the supracrustal rocks and orthogneisses were roughly estimated in eastern Hebei province and in the Taihang-Wutai Mountains area. According to Sun and Wu (1981), the proportion of orthogneisses (their "migmatitic granite") in these two areas is 39 and 42%, respectively, and the supracrustal rock types (amphibolites, metavolcanics, "granulitites", iron formations, and marbles) total about 25% in eastern Hebei and about 30% in the Taihang-Wutai area. In a later study, Wang et al. (1990) state that about one-third of the rocks of the high-grade Archean terrane at Qianan in eastern Hebei province have supracrustal protoliths.

A brief comparison of the Archean geology in northeastern China with that of the Archean greenstone-belt terranes on other cratons is necessary in order to judge the applicability of metallogenetic models for greenstone-belt gold deposits to the Chinese examples (see also Chap. 6). The Archean terrane in eastern Hebei province resembles the "granulite-gneiss belts" of Windley (1984), as exposed in Greenland, Labrador, Scotland, and Limpopo, much more than it does the greenstone belts. Wang et al. (1990) also emphasized the similarity of the Qianxi Group supracrustals (eastern

Hebei province) with the Isua supracrustals of Greenland. The characteristic rock types include marbles, quartzites, possible bimodal gneiss-amphibolite suites, abundant banded iron formations, and pelitic schists. Windley (1984) emphasized that the contrast between the high-grade granulite-gneiss terranes and the greenstone belts is not simply a question of metamorphic grade because there are major differences in their protoliths. One important distinction between a high-grade greenstone belt and a granulite-gneiss terrane, according to Windley, is that the supracrustal rocks in the greenstone belts are typical of the eugeosynclinal environment, with calc-alkaline volcanics, immature graywacke-turbidite sedimentary rocks and Algoma-type iron formations. The granulite-gneiss terranes, on the other hand, contain tholeiitic basalts and shelf sediments including carbonates, quartzites, and pelitic rocks. The latter series of rock types seems to be dominant in most of the Archean sequences in northeastern China according to the descriptions in Yang et al. (1986). However, in contrast to many of the world's granulite-gneiss terranes, the Chinese Archean contains no anorthosite complexes.

There are, in fact, some typical greenstone belt rocks in the Archean of northeastern China but they are very much underrepresented in comparison with the high-grade terranes. Zhai et al. (1985) reported, for example, greenstone/granite-type rocks in the Archean Qingyuan terrane of northern Liaoning province with basal units of mafic and ultramafic rocks overlain by calc-alkaline metavolcanics and eugeosynclinal metasediments, and surrounded by diapric granites.

2.2.1.1 Lithologies in Eastern Hebei Province

It is unfortunate that two different systems of subdivision and nomenclature are currently in use for the Archean and Early Proterozoic rocks in eastern Hebei province. These are compared in Table 2.3. The nomenclature of Sun Dazhong and coworkers of the Tianjin Institute of Geology, Chinese Academy of Geological Sciences (Sun 1984; Sun and Lu 1985) is best known in international publications. However, the authors of this text decided, after some debate, to use the alternative classification of the Ministry of Metallurgical Industry outlined in Table 2.3 because it is more consistent with the usage in the gold districts of eastern Hebei province.

Both systems of nomenclature agree on the subdivision of the Middle- and Late Proterozoic. There are major differences in the terminology of the Early Proterozoic and Archean units and, unfortunately, it is not possible to offer an exact correlation between the two systems, partly because of large differences in the reported thickness of the various units and partly because of the paucity of radiometric ages. For example, the Archean section (pre-2500 Ma) is about 30 km thick according to Q.S. Zhang et al. (1984), whereas Sun (1984) considers the total Archean section to be on the order of only

Table 2.3. Correlation of Early Precambrian rock units in eastern Hebei province

Sun (1984)				This study	
Jingeryu Fm.	Qingbaikou System		Qingbaikou System	Jingeryu Fm.	
Xiamaling Fm.				Xiamaling Fm.	
Tieling Fm.	Jixian System		Jixian System	Tieling Fm.	
Hongshweizhuang Fm.				Hongshweizhuang Fm.	
Wumishang Fm.				Wumishang Fm.	
Yangzhuang Fm.				Yangzhuang Fm.	
Dahongyu Fm.	Changcheng System		Changcheng System	Gaoyuzhuang Fm.	
				Dahongyo Fm.	
Changzhougou Fm.				Tuangshanzi Fm.	
Boluotai Fm.	Qinglonghe Group			Changlingou Fm.	
				Changzhougou Fm.	
Zhangjiagou Fm.	Shuanshanzi Group		Zhuzhangzi Group	Zhalangzhangzi Fm.	
Xiabaicheng Fm.				Shangbaichengzi Fm.	
Luzhangzi Fm.				Zhangjiagou Fm.	
Ciyushan Fm.				Luzhangzi Fm.	
Sanmendian Fm.	Badaohe Group		Dantazi Group	Sanhedian Fm.	
				Nandianzi Fm.	
Wanzhangzi Fm.				Fenghuanzui Fm.	
Wangchang Fm.				Baimaozi Fm.	
			Qianxi Group	Malanyu Fm.	
Santunying Fm.	Qianxi Group			Santunying Fm.	
Shangchuang Fm.				Shangchuang Fm.	

10 km thick. The Early Proterozoic section, according to Q.S. Zhang et al. (1984), totals about 9.5 km in thickness, whereas the time-equivalent section of Sun and coworkers is about 4.5 km thick. This mismatch in the reported thickness of the Early Precambrian units undoubtedly reflects the intense deformation of these rocks. Indeed, Jahn and coworkers have criticized the

usefulness of treating the Qianxi Group in a stratigraphic sense at all (Jahn and Zhang 1984; Jahn et al. 1987; Liu et al. 1990). Liu et al. (1990) recommend that the term Qianxi Group be abandoned in favor of "Qianxi Complex" because of the heterogeneity of the rocks in terms of both lithology and age. We retain the older "stratigraphic" usage here, but note that the definition of the Archean units in northeastern China is in a state of flux.

The following descriptions of the rocks are based in part on published reports and in part on our own observations in outcrops north of the Luanhe River near the towns of Santunying, Qianxi and Qinglong, and in the mining districts of Niuxinshan, Sanjia, and Jinchangyu.

Early Archean: Supracrustal Rocks and Orthogneisses

The Early Archean rocks in eastern Hebei province include metamorphosed supracrustal rocks assigned to the Qianxi Group and orthogneisses which are intrusive into, and include enclaves of, the Qianxi Group rocks. The orthogneisses are not formally included within the Qianxi Group according to current usage, and are therefore described separately below. The Qianxi Group is divided into three formations in ascending order as follows: Shangchuang, Santunying, and Malanyu Formations (see Table 2.3).

Shangchuang Formation
Rocks of granulite facies predominate in the lowest formation of the Qianxi Group, and both mafic and felsic varieties are common. The mafic granulites are granoblastic, two-pyroxene-quartz-plagioclase rocks with or without minor garnet and hornblende. Felsic granulites generally lack orthopyroxene and comprise clinopyroxene, plagioclase, quartz, and minor garnet. Most granulites show extensive retrograde metamorphism to amphibolite facies assemblages, and in addition, many rocks are migmatized with veins and/or schlieren of granitic mobilizates. Minor rock types in the Shangchuang Formation include ultramafic rocks (hornblendites, hornblende-pyroxenites) which may represent mafic dikes, and banded hornblende-magnetite quartzites or magnetite-rich amphibolites interpreted to be metamorphic iron formations. The magnetite-rich rocks are locally mined for iron ore. Detailed study of the mafic and felsic granulites by Jahn and Zhang (1984) suggests that most of these have an igneous parentage with a tholeiitic and calc-alkaline affinity (Chap. 2.2.1.2).

Santunying Formation
This unit contains amphibolites, granulites, and metasedimentary gneisses. There are fewer granulite-facies rocks in the Santunying Formation than in the Shangchuang Formation. The most common rock types are hornblende-plagioclase amphibolites with or without biotite, banded magnetite quartzites, quartz-magnetite amphibolites, and granoblastic felsic biotite gneisses

(so-called leptites). The Santunying Formation is an important source of iron ore in eastern Hebei province. Metapelitic garnet-sillimanite-biotite-plagioclase gneisses, calc-silicate gneisses, and quartzites occur locally. Graphite schists occur at the locality of Qianzhuang in Qinglong county. Retrograde metamorphic textures and mineral assemblages are common in all of the rocks and migmatisation is extensive.

Malanyu Formation
Granulite-facies rocks are rare in the Malanyu Formation and migmatization is not widespread. The main rock types are layered amphibolites, garnet amphibolites, banded pyroxene-hornblende-plagioclase gneisses and schists, and feldspathic biotite schists. Lenses of magnetite-rich amphibolite and magnetite quartzite are also present.

Orthogneisses
Orthogneisses of Archean age in eastern Hebei province are concentrated in the area around Qianan, south of the Luanhe River, and in the easternmost part of the province near the cities of Qinghuangdao and Shanhaiguan (see Fig. 1.3). Orthogneisses are far less common in the area where gold deposits are concentrated, i.e., north of the Luanhe River and west of the Shanhaiguan area. This irregular distribution of orthogneisses and supracrustal rocks has led some workers to divide the Archean terrane of eastern Hebei into two geologic domains (R.M. Wang et al. 1985; Jahn et al. 1987). The supracrustal domain north of the Luanhe River is referred to as the Qianxi highly folded domain or Zunhua Belt, and the area south of the Luanhe River is termed the Qianan Complex or the Qianan Gneiss Dome. These workers did not extend their classification to the Shanhaiguan area, which would logically constitute a third geologic domain.
The orthogneisses of the Qianan area have been described by K.Y. Wang et al. (1985) and by Liu et al. (1990). According to these authors, the gneisses comprise two main compositional and age groups. The first group includes tonalitic to granodioritic gneisses, showing clear intrusive relationships with the supracrustal rocks of the Qianxi Group. These rocks have been metamorphosed together with their supracrustal enclaves at upper amphibolite or granulite facies conditions. The second group of Archean orthogneisses comprises weakly foliated charnockites and granites which intrude both the early orthogneisses and the Qianxi Group metamorphic rocks. The composition of these rocks, summarized in Chapter 2.2.1.2, ranges from tonalitic to granitic, but is significantly more siliceous than the early orthogneisses. K-metasomatism of the country rocks is often observed around the Late Archean intrusions.
The orthogneisses of the Shanhaiguan area have been divided into three intrusive complexes, namely the Anziling Migmatitic Granite, the Jialingkou Quartz Diorite, and the Shanhaiguan Granite. These complexes, especially the Anziling Migmatitic Granite, contain supracrustal relics including iron

formations. Field observations and isotopic age dating indicate that the orthogneisses intruded after the regional metamorphism of the Dantazi Group, i.e., in the Late Archean or Early Proterozoic (Wang et al. 1987).

Late Archean: the Dantazi Group

The Late Archean rocks in eastern Hebei province are assigned to the Dantazi Group. The rock types of the different formations in the Dantazi Group are summarized below in ascending order. Correlations with the nomenclature of Sun (1984) are shown in Table 2.3.

Baimiaozi Formation
This formation contains mainly granoblastic biotite-plagioclase gneisses, hornblende-biotite-plagioclase gneisses and, in some districts, layers of actinolite-magnetite quartzite and cummingtonite-grunerite-magnetite quartzite. The latter locally forms an important source of iron ore.

Fenghuanzui Formation
The Fenghuangzui Formation contains mainly amphibolites with minor hornblende-plagioclase gneisses, leptites with marble lenses, and magnetite-rich hornblende quartzites.

Nandianzi Formation
The Nandianzi Formation consists of biotite-plagioclase gneisses, biotite and chlorite schists, amphibolites, and marble lenses.

2.2.1.2 Chemical Composition and Protolith Rock Types

Qianxi Group
The chemical compositions of meta-igneous rock types of the Qianxi Group (amphibolites, mafic and intermediate-felsic granulites) have been reported by Jahn and Zhang (1984), R.M. Wang et al. (1985), and Wang et al. (1990). According to the studies cited above, the mafic rocks have major element and REE compositions similar to both continental tholeiite basalts and island arc tholeiites. The intermediate and felsic rocks have a calc-alkaline affinity. However, attempts to characterize these rocks in terms of volcanic series and tectonic setting are hampered by the uncertainty introduced by the effects of later geologic events, and any conclusions should be considered with caution. Some aspects of the composition of these rocks taken from the literature are shown in discrimination diagrams in Fig. 2.2.

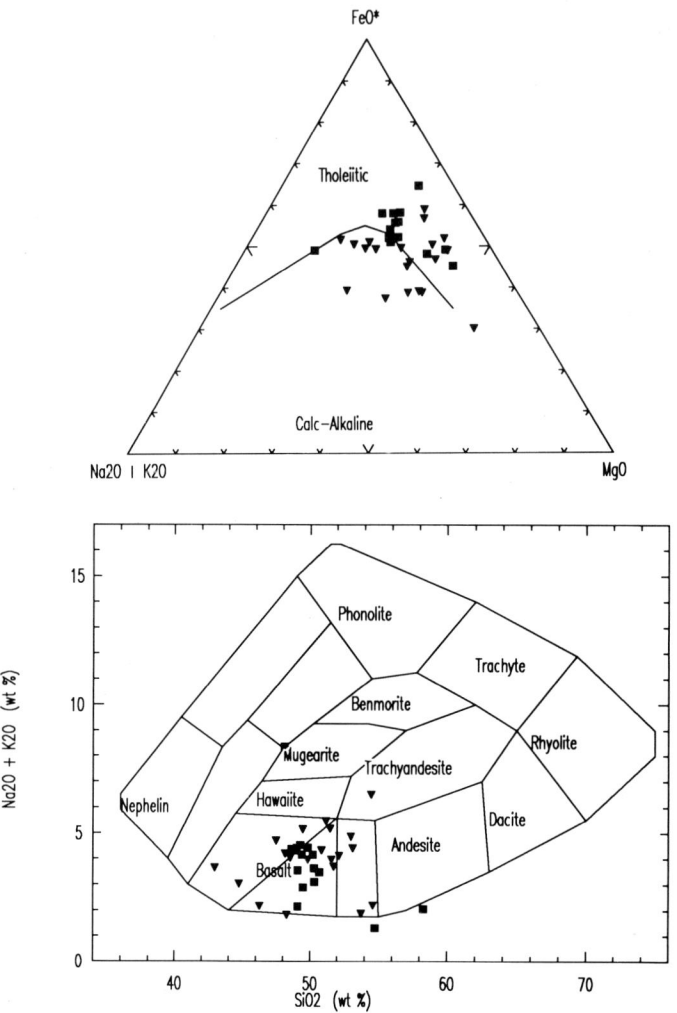

Fig. 2.2. AFM and alkali-silica diagrams showing the compositional range of amphibolites (*triangles*) and mafic granulites (*squares*) from the Qianxi Group. The data plotted were taken from Jahn and Zhang (1984) and Wang et al. (1990)

Orthogneisses

The chemical composition of orthogneisses of the Qianan area have been documented by K.Y. Wang et al. (1985) and by Liu et al. (1990). The early generation of tonalite-granodiorite orthogneisses have major and trace element compositions similar to the calc-alkaline trondhjemite-tonalite-granodiorite (TTG) gneisses typical of other Archean high-grade regions (Wang et al. 1990). The normative feldspar compositions of these rocks are shown in Fig. 2.3. The charnockites have compositions similar to the

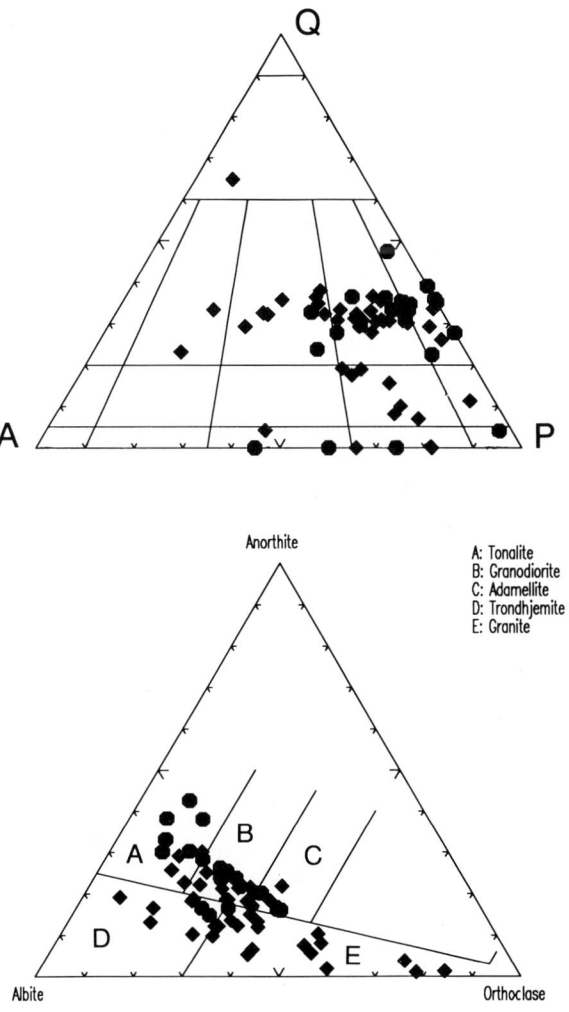

Fig. 2.3. Streckeisen QAP and normative feldspar diagrams showing the compositional range of Archean orthogneisses (*octagons*) and charnockites (*diamonds*) in eastern Hebei province. The data plotted were taken from K.Y. Wang et al. (1985) and Wang et al. (1990)

TTG-orthogneisses, although their bulk compositions tend to be more granitic. Although they formed at different times (see below), both series of orthogneisses and charnockites are thought to have most likely formed by partial melting of the mafic granulite and amphibolite basement (R.M. Wang et al. 1985).

33

Dantazi Group
There are no compositional data available for the rocks of the Dantazi Group, and none were obtained in our investigations. According to the rock types present, the protoliths are interpreted to be mafic and intermediate volcanic rocks in the lower part, grading upward to dominantly intermediate volcano-sedimentary and sedimentary rocks in the upper part (Yang et al. 1986).

2.2.1.3 Isotopic Age

The oldest rocks of the Qianxi Group presently known are supracrustal enclaves in orthogneisses in the Qianan area. The oldest age is a $^{207}Pb/^{206}Pb$ single zircon model age of 3.65 Ga obtained by Liu et al. (1990) from a fuchsite-bearing quartzite. This clearly suggests the existence of a sialic basement older than 3.6 Ga in northeastern China. Amphibolite enclaves in the Qianan orthogneisses have yielded ages of 3.4–3.6 Ga by the Sm-Nd method (Jahn and Zhang 1984; Huang et al. 1986; Jahn et al. 1987). These Sm-Nd ages are interpreted as the crystallization ages of the protoliths. Interestingly, the amphibolites and granulites from the Qianxi Group north of the Luanhe River have not yielded any reliable ages older than 3.0 Ga, which perhaps strengthens the argument that two different Archean domains are present in eastern Hebei. The granulites in the Shangchuang formation of the Qianxi Group north of the Luanhe River have been dated by Rb-Sr, Sm-Nd, and zircon U-Pb methods by different laboratories and all give ages around 2.5 Ga (Pidgeon 1980; Jahn and Zhang 1984). Jahn and Zhang (1984) concluded from their Sm-Nd and Rb-Sr data that the 2.5 Ga ages reflect the age of granulite-facies metamorphism, but that the protolith basalts cannot have formed much more than 100 Ma prior to metamorphism. The orthogneisses intrusive into the Qianxi Group in the Qianan area have been dated by U-Pb and Pb-Pb methods on zircon, as summarized by Liu et al. (1990). The results show a range of ages between about 2.9 and 2.5 Ga with a pronounced cluster at about 2.5 Ga. The ages were all interpreted as crystallization ages. K.Y. Wang et al. (1985) reported a Rb-Sr whole rock isochron age of 2650 ± 50 Ma from charnockites from the same area, which they also interpreted as the crystallization age of the rocks. The orthogneisses in the Shanhaiguan area seem to be younger based on their Rb-Sr ages of 2412–2446 Ma (Wang et al. 1987).

Figure 2.4 summarizes the published isotopic age data from the Archean rocks in eastern Hebei. The earliest supracrustal rocks formed at about 3.6 Ga and these rocks preserve evidence of a still older sialic basement. Extensive intrusion of tonalite-trondjhemite rocks at about 2.5 Ga in the Qianan area was broadly contemporaneous with granulite facies regional metamorphism. The 2.5 Ga events, referred to as the Fuping Orogeny (Chap. 2.1.2), are reflected in isotopic ages from other areas of northeastern

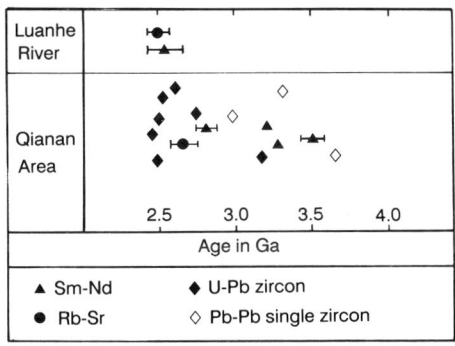

Fig. 2.4. Summary of isotopic age data from Archean rocks in eastern Hebei province. (After Liu et al. 1990)

China as well. To date, no good evidence of Early Archean rocks (pre-3.0 Ga) in notheastern China outside of eastern Hebei province has been reported. A younger group of ages from the Qianxi Group rocks, not shown on Fig. 2.4, represents later thermal events related to the Wutai (2 Ga) and/or Zhongtiao (1.7 Ga) orogenies. Huang et al. (1986) report a Sm-Nd whole rock isochron age of 1.7 Ga for a quartz diorite whose depleted mantle model age is 3.4 Ga. Jahn and Zhang (1984) obtained a Rb-Sr mineral isochron age of 1.68 Ga from granulites whose whole-rock age is 2.5 Ga. K.Y. Wang et al. (1985) attributed the Rb-Sr whole rock age of 2.1 Ga for gray gneisses near Qianan to isotopic resetting by the Wutai Orogeny.

Relatively few radiometric age data are available from the Dantazi Group rocks. Yang et al. (1986) cite K-Ar hornblende and whole-rock ages ranging from 2435 to 2660 Ma and one Rb-Sr whole rock age of 2523 ± 139 Ma for the Dantazi Group as a whole. Sun and Lu (1985) reported age data from the correlative rocks of the Badaohe Group (see Table 2.3), including a Rb-Sr whole rock isochron age of 2550 ± 45 Ma from the Wangcheng Formation and an U-Pb zircon age of 2494 ± 24 Ma from migmatitic granite in the Badaohe Group. These were interpreted by Sun and Lu (1985) as the age of high-grade metamorphism and migmatization related to the Fuping Orogeny.

2.2.1.4 Metamorphism

The P-T conditions of granulite metamorphism in rocks of the Qianxi Group have been estimated at 750 to 850 °C and 8 to 11 kbar based on two-pyroxene and garnet-pyroxene equilibria (Zhang and Cong 1982; Sun 1984; Sills et al. 1987a). The presence of sillimanite, cordierite, and garnet in

35

metapelitic horizons in the Qianxi Group suggests medium-pressure metamorphic conditions. Retrograde metamorphism is a common feature of the Qianxi Group rocks, and many of the granulite facies rocks have been transformed into retrograde amphibolites. Sun (1984) reported P-T estimates of retrograde metamorphism at 550–750 °C and 6–10 kbar. R.M. Wang et al. (1985) reported garnet-biotite temperatures of about 600 °C from a retrograded granulite from the Qianan area.

Quantitative estimates of metamorphic conditions in the rocks of the Dantazi Group have not been published. The mineral assemblages of the rocks suggest that the metamorphism reached the upper amphibolite facies and the onset of anatexis. No reliable estimates of pressure can be made from the available information.

2.2.2 Proterozoic Rocks

The Proterozoic Eon in China is defined by Yang et al. (1986) as extending from 2500 to 850 Ma. Between 850 Ma and the beginning of the Cambrian Period Chinese geologists recognize the "Sinian System". The Proterozoic rocks in eastern Hebei province can be usefully divided into two series based on their structural and metamorphic nature. The first is of Early Proterozoic age (ca. 2500–1800 Ma), and it forms a strongly folded and foliated medium-grade metamorphic basement which shares some tectonic elements with the Archean. The second series formed in the Middle and Late Proterozoic (ca. 1800–850 Ma), and it consists of platformal clastic and carbonate sedimentary sequences which are weakly metamorphosed and only gently folded.

The Proterozoic rocks in northeastern China have been much less intensely studied from a geochemical and isotopic standpoint than have the Archean rocks. It is not possible to provide details of the chemical composition, isotopic composition, and P-T conditions of metamorphism as was the case for the Archean rocks. For the present purpose, this lack of information is not critical, since most of the gold deposits in eastern Hebei occur in the Archean rocks. The discussion below is therefore limited to a listing of the Proterozoic formations present in eastern Hebei province and their lithologic makeup, together with such age data as are available. A full discussion of Proterozoic stratigraphy in China is given in Yang et al. (1986) and references therein.

The following lithologic descriptions and age data are taken mostly from information in Yang et al. (1986), Sun and Lu (1985), and partly from unpublished data from the Ministry of Metallurgical Industry. The stratigraphic nomenclature is given in Table 2.3.

2.2.2.1 Early Proterozoic

The Early Proterozoic rocks are poorly represented in eastern Hebei province in contrast to the Archean rocks. In many parts of eastern Hebei province Middle- to Late Proterozoic rocks rest unconformably or tectonically on the Archean, and the Early Proterozoic is entirely missing (see Fig. 1.3). The best-exposed type section of the Early Proterozoic in northeastern China is found in the Wutai–Taihang mountains area west of Beijing (area VI in Fig. 2.1). In this area, the Early Proterozoic (Wutai Group) is characterized by greenschist or lower amphibolite facies flysch-like turbidite sequences with minor BIF layers, and intermediate to felsic volcanoclastic rocks (Yang et al. 1986). Similar rock types occur in eastern Hebei province but their nomenclature is different (see Table 2.2), and correlation between the two areas is tentative.

The Early Proterozoic rocks in eastern Hebei province are represented by the Zhuzhangzi Group, which is exposed in a limited area east of Qinglong (see Fig. 1.3). Five formations of the Zhuzhangzi Group are recognized as described below in ascending order:

Sanhedian Formation
The formation consists of feldspathic sericite-quartzites, sericite-quartz schists, sericite schists, and magnetite-rich quartzites.

Luzhangzi Formation
The dominant rock types include amphibolites with relict pillow structure, hornblende schists, and chlorite-hornblende schists. Isotopic ages (Rb-Sr) reported from the unit are 2217 ± 43 Ma (Lu and Huang 1987) and 2193 Ma (Sun 1984).

Zhangjiagou Formation
The main rock types are meta-conglomerates with volcanogenic clasts and biotite leptites. The isotopic age (Rb-Sr) is 2223 ± 136 Ma (Lu and Huang 1987). Sun (1984) reported a Rb-Sr age of 2398 Ma.

Shangbaichengzi Formation
This formation contains biotite leptites, biotite-hornblende leptites, hornblende schists, and lenses of iron-rich quartzite showing cross-bedding. The isotopic age (Rb-Sr) is 1824 ± 66 Ma (Lu and Huang 1987).

Zhalangzhangzi Formation
This unit includes sericite schists, biotite schists, biotite leptites, and iron-rich quartzites, all locally containing garnet.

2.2.2.2 Middle to Late Proterozoic

The Middle- and Late Proterozoic rocks (1800–850 Ma) are well represented in eastern Hebei province, and one of best exposed type sections in China, some 9.5 km thick, is at Jixian, located a few tens of kilometers west of Qianan (for location, see Fig. 1.3). The rocks consist of weakly deformed, very low-grade metamorphic sedimentary and volcano-sedimentary sequences. The Middle- and Late Proterozoic rocks are divided into three systems; in ascending order these are the Changcheng System, Jixian System, and Qingbaikou System (Table 2.3). A brief description of the formations in each system is given below.

1. The Changcheng System

Changzhougou Formation
The basal formation of the Changcheng system consists of conglomerates and coarse sandstones in the lower part and sandy siltstones in the upper part.

Changlingou Formation
This unit consists of silty slates and shales in the lower and middle parts; and slates, shales and minor carbonaceous dolomites in the upper part. The isotopic age (K-Ar) is 1875–1817 Ma (Zhong 1975).

Tuangshanzi Formation
This unit consists of argillaceous and siliceous dolomites in the lower part, and silty micrite dolomites and dolomitic sandstones in the upper part. The isotopic age is 1776 Ma (U-Pb whole rock, Zhong 1975).

Dahongyu Formation
This formation contains transgressive sandstones and arkoses, flinty micrites and siliceous laminated dolomites. Potassic volcanic rocks occur in the middle part. The isotopic age is 1678 Ma (K-Ar glauconite, Zhong 1975).

Gaoyuzhuang Formation
The formation contains basal quartz arenites, dolomitic sandstones in the lower part, flinty dolomites in the middle part, and an upper part of manganiferous dolomites and micritic dolomitic limestones.

2. The Jixian System

Yangzhuang Formation
The main rock types are silty argillaceous dolomites with dolomitic limestones, and limestones.

Wumishan Formation
The main rock types are dolomites and flinty banded dolomites in the lower part, and dolomitic limestones in the upper part.

Hongshweizhuang Formation
The unit contains dolomites in the lower part, and slates with thin beds of sandstone in the upper part.

Tieling Formation
This formation contains dolomitic limestones, manganiferous limestones and shales in the lower part, and stromatolitic limestones in the upper part. The isotopic age is 1205 to 1132 Ma (K-Ar glauconite, Zhong 1975).

3. The Qingbaikou System

Xiamaling Formation
The Xiamaling Formation contains conglomerates, ferruginous sandstones, and paleosols at the base, which represent an erosional surface on the underlying limestones. The basal part is succeeded by shales and sandstones in the lower part, and shales with siltstones in the upper part.

Jineryu Formation
This formation forms a transgressive sequence with feldspathic conglomerates at the base followed upward by arkosic sandstones and shales in the lower part and dolomitic micritic limestones in the upper part. Yang et al. (1986) reported a K-Ar glauconite age of 899–855 Ma for the formation.

2.2.2.3 Metamorphism and Nature of the Protoliths

The metamorphic grade of the Zhuzhangzi Group rocks is in the greenschist to epidote-amphibolite facies according to the reported mineral assemblages. The protoliths are dominated by volcanic and volcaniclastic rocks, with mafic volcanic agglomerates and lavas at the base, overlain by intermediate to felsic volcanic rocks with upward-increasing volcaniclastic components. The protoliths of the upper formations are dominated by volcaniclastic rocks intercalated with semipelitic sedimentary rocks.
The Middle and Late Proterozoic rocks are essentially unmetamorphosed epicontinental platform sequences representing cyclic sedimentation of clastic and carbonate material. Volcanogenic material is locally present but is rare in comparison to the Early Proterozoic section.

2.3 Structural Geology

Eastern Hebei province is situated in the eastern part of the Sino-Korean platform near the border between two "second-order" tectonic units, namely the Inner Mongolian Axis and the Yanshan Platformal Fold Belt. These are marked in Fig. 1.1 by the letters A and B, respectively. The area has been subject to repeated orogenies, as summarized in Chapter 2.1, the most intense of which were the Fuping (2500 Ma), Zhongtiao (1800 Ma), and Yanshan (200–150 Ma). Particularly the Yanshan Orogeny produced extensive, deep fracture/fault zones, some of crustal scale.

2.3.1 Precambrian Structures

Figure 2.5 shows a simplified tectonic map of Early Precambrian exposures in eastern Hebei province after Sun et al. (1989). The Early Precambrian rocks are exposed in a series of generally E–W-trending anticlinoria. These relatively open folds represent a late stage of folding attributed by most authors to the Zhongtiao Orogeny (ca. 1800 Ma). Folds of an earlier stage (Fuping and pre-Fuping) are tight to isoclinal with nearly N–S axes and west-dipping axial planes (Sun 1984; Sun et al. 1989). As a result of the late stage folding, the strike direction of lineations and foliations in the Archean and Early Proterozoic rocks changes from dominantly NE–SW in the north to N–S in the south.

According to Sun (1984), the area shown in Fig. 2.5 is divided into an eastern and a western tectonic unit by the major NW-trending Lengkou Fracture Zone (marked on the tectonic map). The western unit is uplifted relative to the eastern, thus the Archean rocks are most abundant west of the Lengkou Fracture Zone. In the western tectonic unit early isoclinal folds are well exposed in the Qianxi Group in the Santunying-Taipingzhai region of Qianxi county, at Malanyu in the western part of Zunhua county, and in the western part of Qianan county. Late stage open folding is best represented by the large Malanyu-Taipingzhai Anticlinorium (MTA on the map) which extends E–W through the center of the area.

In the eastern tectonic unit Early Proterozoic rocks are more commonly exposed and there are large areas of migmatitic granite. The early stage of tight folds with N–S axial planes dipping west is present, but the early folds are more open than in the western tectonic unit. On the other hand, the E–W-trending late stage folds are tighter than in the western tectonic unit; an example shown in the tectonic map is the Qianzhangzi-Longwangmiao Anticlinorium (QLA on the map).

The Archean and Proterozoic structures played a major role in influencing later structures, especially the position of intrusions and main fracture zones, both of which are important for localizing gold mineralization. Both the early and Late Precambrian fold generations developed reverse faults in

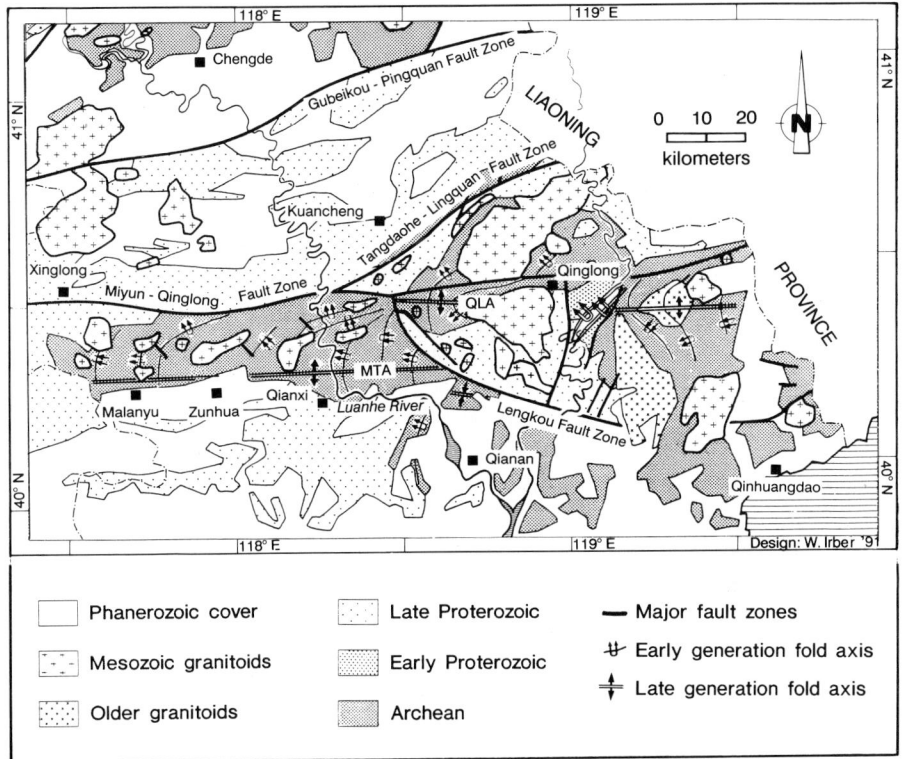

Fig. 2.5. Geologic map of part of eastern Hebei province showing the main structural elements from Sun et al. (1989) discussed in the text

their hinge zones which are represented by ductile shear zones. These facilitated later tectonic movements and magmatic intrusions.

2.3.2 Regional Faults and Lineaments

The most important structural features in northeastern China in terms of gold metallogeny are the faults and fracture zones because they influence the distribution of both magmatic intrusions and hydrothermal circulation. Figure 2.6 shows the most important fault zones in eastern Hebei province and adjacent areas based on geologic and geophysical surveys, superimposed on a lineament interpretation map derived from 1:250 000 scale LANDSAT images. The rose diagram in the inset shows the dominant NE-SW direction of lineaments and a significant component with N-S and lesser component with E-W trends. These lineaments coincide with known faults and joint patterns measured on the ground, and the mineralized

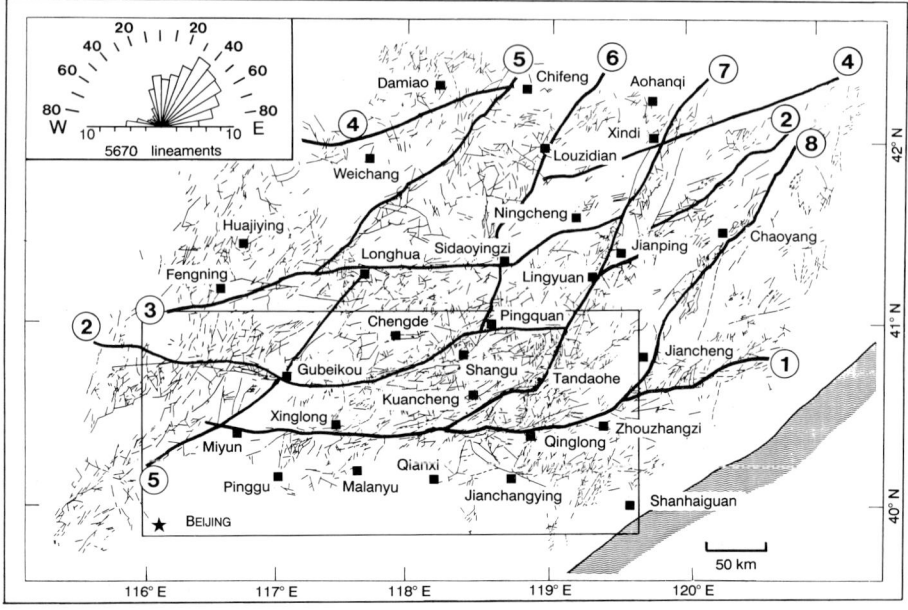

Fig. 2.6. Map showing the regional distribution of lineaments from LANDSAT images of eastern Hebei province and adjacent parts of Liaoning province and southern Inner Mongolia. The numbered fault zones (*heavy lines*) are described in the text. The *boxed area* corresponds to the geologic map of Fig. 1.3

structures in the mining districts also show the dominant NE–SW orientation, as discussed in Chapter 3.

2.3.3 Structural Features of the Major Fault Zones

The area of eastern Hebei province is dominated by NE–SW, N–S, and E–W-trending faults. Multiple faults of a given direction tend to form at regular spacing so that the intersection of the three directions produces rhombic structural blocks. The fault zones are of primary importance in controlling the distribution of gold mineralization both in terms of the location of mining districts and of the associated magmatic intrusions. On a scale of single mining districts the mineralization forms along secondary faults and not directly along the major fault zones.

In the following sections the most important deep fault zones are described in terms of their relative age, orientation, and sense of movement. The fault zones are divided into three groups according to their dominant orientation, namely E–W, NE–SW, and NW–SE, and described in sequence from older to younger. The number given before the name of each fault zone refers to the corresponding number on the map of Fig. 2.6.

2.3.3.1 E–W Fault Zones

The E–W fault zones are developed particularly well in the area south of Longhua. From south to north four major fault zones are recognized:

1. Miyun-Qinglong Fault Zone

This fault zone has a length of more than 300 km. If the secondary faults on either side of the main zone are included, the total width is 10 km. The angle of dip varies, but is generally steep and to the south. The zone is cut by some faults of NE, NNE, and N–S trend, showing that it is relatively early. The geologic nature of the fault zone changes along its length. Most parts of the fault show a reverse sense of movement. Where it separates the Archean gneisses from the Middle to Late Proterozoic rocks, the fault zone shows abundant evidence of ductile deformation including mylonite and associated recumbent folds. The fault zone splays and locally pinches out in the eastern part of the area due to the influence of NNE-trending structures. The overall trend also changes from E–W to ENE in this area.

Most of the Archean exposures in eastern Hebei province are found south of the Miyun-Qinglong fault zone, along the Malanyu-Taipingzhai Anticlinorium (MTA in Fig. 2.5). Numerous intrusive bodies and associated gold mineralizations are distributed along the fault zone.

2. Gubeikou-Pingquan Fault Zone

This zone consists of groups of medium scale faults with E–W and ENE–WSW direction. The faults cut all units from Archean gneisses to Jurassic volcanics. The fault zone is cut by later faults of NW, NNE, and N–S trend, showing that it is relatively early. The western part of the fault zone strikes E–W and dips to the south. The middle part of the zone consists of a series of roughly parallel minor faults and fractures with a NE–SW trend which merge locally with major NE–SW structures. The eastern part of the zone is offset from the other parts by a younger NE-trending fault zone (number 7, Fig. 2.6). In this area the zone consists of an en-echelon series of NE-trending parallel faults. In the Lingyuan district, the fault zone shows reverse movement with the Archean Jianping Group thrust over the Middle and Late Proterozoic rocks.

A marked difference in rock types occurs across the Gubeikou-Pingquan fault zone, which led many geologists to consider that the zone marks the contact between the Inner Mongolian Axis and the Yanshan Fold Belt. Intense vertical movements of the Inner Mongolian Axis relative to the Yanshan Fold Belt during the Yanshan Orogeny led to the intrusion of a zone of mafic-ultramafic rocks along the fault zone at deep crustal levels. Some parts of this zone are related to gold mineralization, notably at Gubeikou, Pingquan, and in the east at Jianping.

3. Fengning-Ningcheng Fault Zone

This fault zone (referred to as the Fengning-Longhua zone in Ren et al. 1987) consists of a major fault and multiple small E–W-trending faults. In many places the zone is cut by NNE- and NW-trending faults. The eastern part of the zone terminates on an intensive major NNE-trending fault zone (number 7, Fig. 2.6). The dip angle of most faults in this zone is near vertical; east of Longhua the faults dip to the north and west of Longhua they dip to the south. The fault zone is most prominently developed in the Archean Dantazi Group. Southeast of Longhua, for example, a cataclastic zone 30–40 m wide is developed. Horizontal slickensides show that the movement involved strike-slip components. According to Ren et al. (1987), the main period of movement of the Fengning-Longhua fault zone was in the Proterozoic.

Many granitic and granodioritic intrusive bodies, most of Yanshanian age, are distributed along both sides of the fault zone. Gold mineralization is particularly abundant at the intersection of this zone with NE-trending fault zones, as especially evident in the Longhua district.

4. Kangbao-Chifeng Fault Zone

The Kangbao-Chifeng zone forms the northern boundary of the Inner Mongolian Axis. The zone does not crop out clearly due to widespread Mesozoic cover, and is not well expressed on satellite imagery. Geologic evidence for a major fault lies in the lithologic contrast across the zone. To the north, Late Paleozoic strata are common, while in the south the Paleozoic is absent and Archean gneisses are widespread. Furthermore, Tertiary basalt and Paleozoic ultramafic intrusions occur along the fault zone. The nature of the fault zone is compressive (reverse movement). The zone is offset by NNE- and NNW-trending faults. Gold mineralization related to the zone occurs mainly where it intersects NNE-trending faults.

2.3.3.2 NE–SW Fault Zones

The NE–SW-directed faults are the most abundant in the area. These are generally younger than the E–W-trending fault zones described above and locally offset them. The most important of the NE–SW-trending fault zones are described below. The numbers refer again to Fig. 2.6.

5. Miyun-Longhua-Mijiayingzi Fault Zone

The fault zone is divided into two separate sections at Longhua (Fig. 2.6). The northern section shows reverse and strike-slip movement where it cuts Jurassic strata. The southern section cuts two major E–W-trending fault zones. The fault zone is interpreted to have formed during the Late Yanshan Orogeny because isotopically dated granites of that age occur along the fault and because it clearly cuts across the earlier E–W-directed fault zones. Gold